Principles of Compiler Design

ITL Education Solutions Limited
Research and Development Wing
New Delhi

ISBN 978-81-317-6126-7

First Impression, 2012
Second Impression,
Third Impression, 2013

Published by Dorling Kindersley (India) Pvt. Ltd, licensees of Pearson Education in South Asia.

Head Office: 7th Floor, Knowledge Boulevard, A-8(A), Sector 62, Noida 201 309, UP, India.
Registered Office: 11 Community Centre, Panchsheel Park, New Delhi 110 017, India.

Compositor: MacroTex Solutions
Printer: Sai Print O Pack Pvt. Ltd.

Contents

Preface

A compiler is a program that translates high-level languages such as C, C++ and Java into lower-level languages like equivalent machine codes. These machine codes can be understood and directly executed by the computer system to perform various tasks. Given its importance, Compiler Design is a compulsory course for B.Tech. (CSE and IT) students in most universities. The book in your hand *Principles of Compiler Design* in its unique easy-to-understand question-and-answer format directly addresses the need of students enrolled in these courses.

The questions and corresponding answers in this book have been designed and selected to cover all the basic and advanced level concepts of Compiler Design including lexical analysis, syntax analysis, code optimization and generation, and error handling and recovery. This book is specifically designed to help those who are attempting to learn Compiler Design by them. The organized and accessible format allows students to quickly find the questions on specific topics.

The book *Principles of Compiler Design* forms a part of series called the *Express Learning Series,* which has a number of books designed as quick reference guides.

Unique Features

1. Designed as student friendly self-learning guide. The book is written in a clear, concise and lucid manner.
2. Easy-to-understand question-and-answer format.
3. Includes previously asked as well as new questions organized in chapters.
4. All types of questions including multiple-choice questions, short and long questions are covered.
5. Solutions to the numerical questions asked in the examinations are provided.
6. All ideas and concepts are presented with clear examples.
7. Text is well structured and well supported with suitable diagrams.
8. Inter-chapter dependencies are kept to a minimum.

Chapter Organization

All the question–answers are organized into 11 chapters. The outline of the chapters are as follows:

- *Chapter 1* provides an overview of compilers. It discusses the difference between interpreter and compiler, various phases in the compilation process with the help of an example, error-handling in compilers and the concept of cross compiler and bootstrapping. This chapter forms the basis for the rest of the book.

- *Chapter 2* details the lexical analysis phase including lexical analyzer, tokens, patterns and lexemes, strings and languages and the role of input buffering. It also explains regular expressions, transition diagrams, finite automata and the design of lexical analyzer generator (LEX).
- *Chapter 3* describes the context free grammars (CFG) along with its ambiguities, advantages and capabilities. It also discusses the difference between regular expressions, and CFG and introduces context free language.
- *Chapter 4* spells out the syntax analysis phase including role of parser, categories of parsing techniques and parsed tree. It elaborates the top–down parsing techniques, which include backtracking and non-backtracking parsing techniques.
- *Chapter 5* deals with bottom up parsing techniques, which include simple LR (SLR) parsing, canonical LR (CLR) parsing and lookahead LR (LALR) parsing. The chapter also introduces the tool YACC to show the automatic generation of LALR parsers.
- *Chapter 6* explains the concept of syntax-directed translations (SDT) and syntax-directed definitions (SDD).
- *Chapter 7* expounds on how to generate an intermediate code for a typical programming language. It discusses different representations of the intermediate code and also introduces the concept of backpatching.
- *Chapter 8* throws light on type checking process and its rules. It also explains type expressions, static and dynamic type checking, design process of a type checker, type equivalence and type conversions.
- *Chapter 9* familiarizes the reader with runtime environment, its important elements and various issues it deals with. It also discusses static and dynamic allocation, control stack, activation records and register allocation.
- *Chapter 10* explores the usage of symbol table in a compiler. It also discusses the operations performed on the symbol table and various data structures used for implementing the symbol table.
- *Chapter 11* familiarizes the reader with code optimization and the code generation process.

Acknowledgements

- Our publisher Pearson Education, their editorial team and panel reviewers for their valuable contributions toward content enrichment.
- Our technical and editorial consultants for devoting their precious time to improve the quality of the book.
- Our entire research and development team who have put in their sincere efforts to bring out a high-quality book.

Feedback

For any suggestions and comments about this book, please contact us at itlesl@rediffmail.com. Hope you enjoy reading this book as much as we have enjoyed writing it.

Rohit Khurana
Founder and CEO
ITL ESL

1

Introduction to Compilers

1. What do you understand by the terms translator and compiler?

Ans: A **translator** or **language processor** is a program that translates an input program written in a programming language into an equivalent program in another language. **Compiler** is a type of translator, which takes a program written in a high-level programming language as input and translates into an equivalent program in low-level language such as machine language or assembly language. The program written in high-level language is known as **source program**, and the program converted into low-level language is known as **object** (or **target**) **program**. Moreover, compiler traces the errors in the source program and generates the error report. Without compilation, no program written in a high-level language can be executed. After compilation only the program in machine language is loaded into the memory for execution. For every programming language, we have a different compiler; however, the basic tasks performed by every compiler are same.

2. Explain the steps required for the execution of a high-level language program with the help of compiler.

Ans: The execution of a high-level language program is performed basically in two steps:

- **Compilation or translation:** During compilation the source program is translated into the target program. The target program can either be a machine code or an assembly language code. If the target program is executable machine language code, then it can be executed directly to generate the output. Figure 1.1 shows the compilation phase.

Figure 1.1 Compilation of Source Program

- **Execution of the target program:** During execution, the target program is first loaded into the main memory and then the user interacts with the target program to generate the output. The execution phase is shown in Figure 1.2.

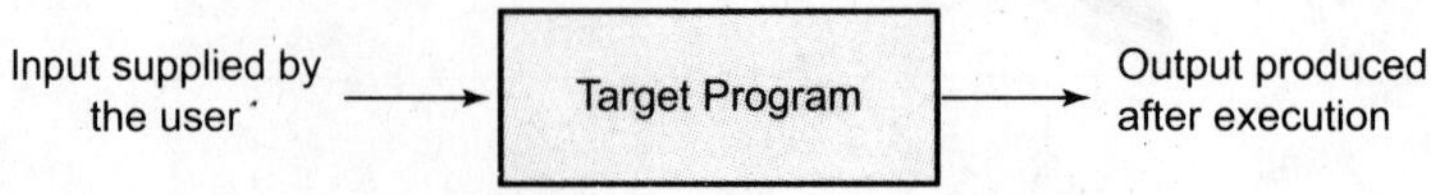

Figure 1.2 Executing Target Program

3. What are the difference between compiler and interpreter?

Ans: Compiler translates the whole source program into the target program in one step (see Figure 1.1). That is, it first scans the entire input program and then translates it into the target program. The target program is then executed separately for generating the output according to the given inputs.

Interpreter, on the other hand, directly executes the source program line by line according to the given inputs. That is, translation and execution of each statement are carried out side by side. Thus, separate execution of the program is not required. The line by line execution of the program provides better debugging environment than a compiler. The main drawback of an interpreter is that the execution time of an interpreted program is generally slower than that of a compiled program because the program needs to be translated every time it is executed. The interpretation process is shown in Figure 1.3.

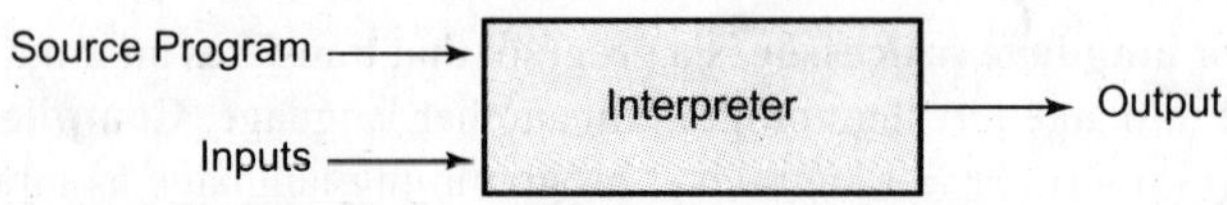

Figure 1.3 Working of an Interpreter

4. What do you understand by the term cousins of compiler?

Ans: The term 'cousins of compiler' refers to the type of programs which are required for the execution of the source program. These are the programs along with which compiler operates. The cousins of compilers are *preprocessors, assemblers,* and *loaders and link editors.*

- **Preprocessors:** Before compilation, the source program is processed by the preprocessor to prepare it for the compilation. The preprocessor program creates modified source program from the original source program by replacing the preprocessor directives with the suitable content. The new source program acts as an input to the compiler (see Figure 1.4). Preprocessor performs various tasks as given here.
  - It permits the user to include the header files in the program and user can make use of the functions defined in these header files.
  - It permits the user to include macros in the program. **Macros** are the small set of instructions that are used in a program repetitively. Macros have two attributes, namely, *macro name* and *macro definition*. Whenever the macro name is encountered in the program then it is replaced by the macro definition (set of statements correspond to the macro).

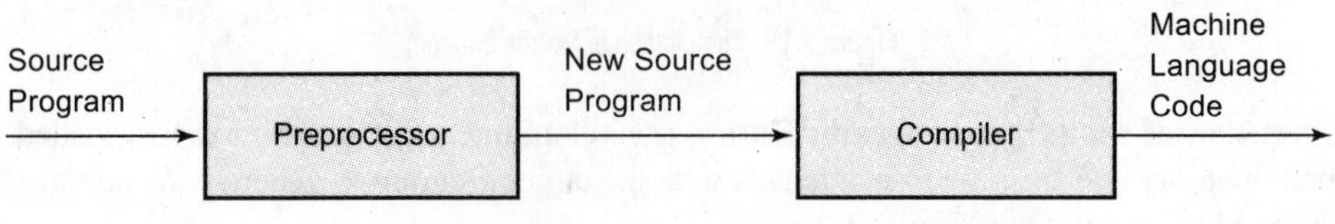

Figure 1.4 Preprocessor's Role

- **Assemblers:** In some cases, compiler generates the target program in assembly language. In that case, the assembly language program is given to the assembler as input. An assembler then translates the assembly language program into machine language program which is relocatable machine code. An assembly language program is in mnemonics.

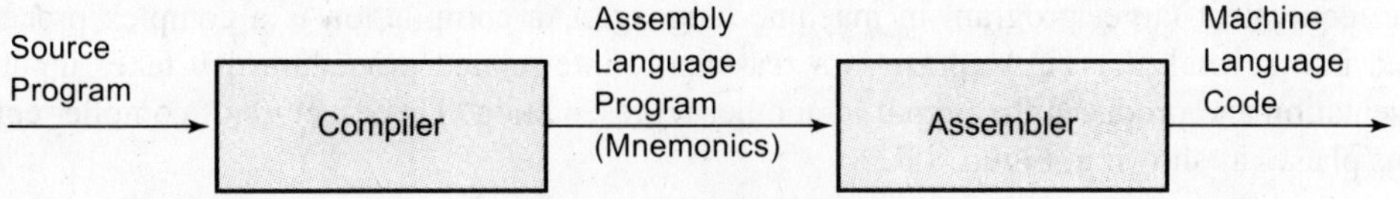

Figure 1.5 Assembler's Role

- **Loaders and link editors:** The larger source programs are compiled in small pieces by the compiler. To run the target machine code of any source program successfully, there is a need to link the relocated machine language code with library files and other relocatable object files. So, loader and link editor programs are used for the link editing and loading of the relocated codes. **Link editors** create a single program from several files of relocated machine code. **Loaders** read the relocated machine code and alter the relocatable addresses. To run the machine language program, the code with altered data and commands is placed at the correct location in the memory.

5. Discuss the steps involved in the analysis of a source program with the help of a block diagram.

Ans: The steps involved in the analysis of source program are given below.

- Source program acts as an input to the preprocessor. Preprocessor modifies the source code by replacing the header files with the suitable content. Output (modified source program) of the preprocessor acts as an input for the compiler.
- Compiler translates the modified source program of high-level language into the target program. If the target program is in machine language, then it can be executed directly. If the target program is in assembly language, then that code is given to the assembler for translation. Assembler translates the assembly language code into the relocatable machine language code.
- Relocatable machine language code acts as an input for the linker and loader. Linker links the relocatable code with the library files and the relocatable objects, and loader loads the integrated code into memory for the execution. The output of the linker and loader is the equivalent machine language code for the source code.

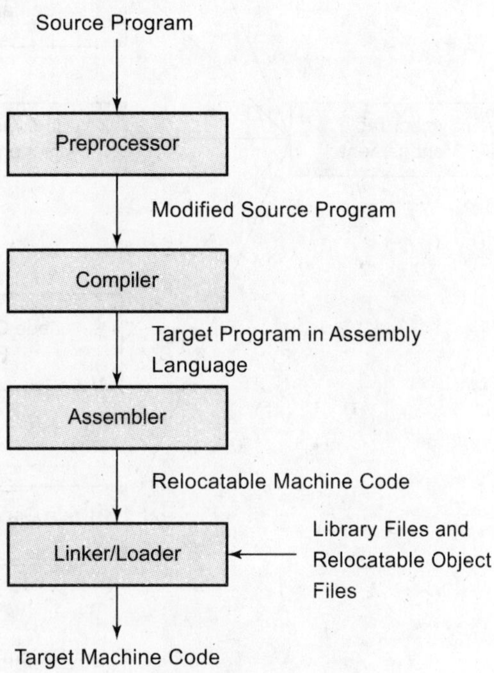

Figure 1.6 Block Diagram of Source Program Analysis

6. Explain the different phases of compiler with diagram.

Or

Explain the structure of compiler.

Ans: Compiler translates an input source program written in any high-level programming language into an equivalent target program in machine language. As compilation is a complex process, it is divided into several phases. A **phase** is a reasonably interrelated procedure that takes input in one representation and produces the output in another representation. The structure of compiler comprises various phases as shown in Figure 1.7.

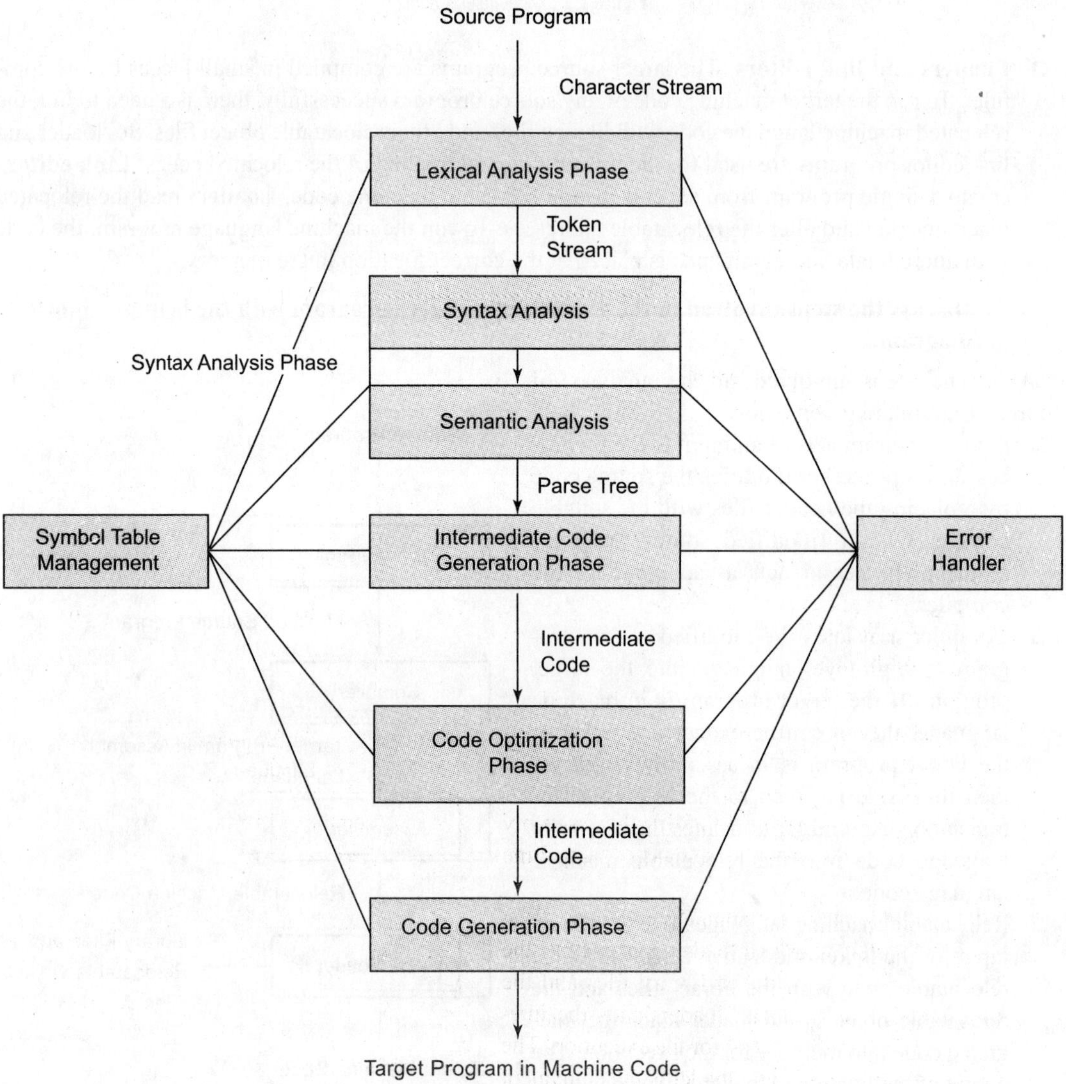

Figure 1.7 Phases of a Compiler

- **Lexical analysis phase:** Lexical analysis (also known as **scanning**) is the first phase of a compiler. Lexical analyzer or scanner reads the source program in the form of character stream and groups the logically related characters together that are known as **lexemes**. For each lexeme, a **token** is generated by the lexical analyzer. A stream of tokens is generated as the output of the lexical analysis phase, which acts as an input for the syntax analysis phase. Tokens can be of different types, namely, *keywords, identifiers, constants, punctuation symbols, operator symbols*, etc. The syntax for any token is:

  ```
  (token_name, value)
  ```

  where `token_name` is the name or symbol which is used during the syntax analysis phase and `value` is the location of that token in the symbol table.
- **Syntax analysis phase:** Syntax analysis phase is also known as **parsing**. Syntax analysis phase can be further divided into two parts, namely, *syntax analysis* and *semantic analysis*.
  - **Syntax analysis:** Parser uses the `token_name` token from the token stream to generate the output in the form of a tree-like structure known as **syntax tree** or **parse tree**. The parse tree illustrates the grammatical structure of the token stream.
  - **Semantic analysis:** Semantic analyzer uses the parse tree and symbol table for checking the semantic consistency of the language definition of the source program. The main function of the semantic analysis is type checking in which semantic analyzer checks whether the operator has the operands of matching type. Semantic analyzer gathers the type information and saves it either in the symbol table or in the parse tree.
- **Intermediate code generation phase:** In intermediate code generation phase, the parse tree representation of the source code is converted into low-level or machine-like intermediate representation. The intermediate code should be easy to generate and easy to translate into machine language. There are several forms for representing the intermediate code. Three address code is the most popular form for representing intermediate code. An example of three address code language is given below.

  ```
  x1 = x2 + id
  id1 = x3
  ```

- **Code optimization phase:** Code optimization phase, which is an optional phase, performs the optimization of the intermediate code. Optimization means making the code shorter and less complex, so that it can execute faster and takes lesser space. The output of the code generation phase is also an intermediate code, which performs the same task as the input code, but requires lesser time and space.
- **Code generation phase:** Code generation phase translates the intermediate code representation of the source program into the target language program. If the target program is in machine language, the code generator produces the target code by assigning registers or memory locations to store variables defined in the program and to hold the intermediate computation results. The machine code produced by the code generation phase can be executed directly on the machine.

Symbol table management: A symbol table is a data structure that is used by the compiler to record and collect information about source program constructs like variable names and all of its attributes, which provide information about the storage space occupied by a variable (name, type, and scope of the variables). A symbol table should be designed in an efficient way so that it permits the compiler to locate the record for each token name quickly and to allow rapid transfer of data from the records.

Error handler: Error handler is invoked whenever any fault occurs in the compilation process of source program.

Both the symbol table management and error handling mechanisms are associated with all phases of the compiler.

7. **Discuss the action taken by every phase of compiler on the following instruction of source program while compilation.**

`Total = number1 + number2 * 5`

Ans: Consider the source program as a stream of characters.

```
Total = number1 + number2 * 5
```

- **Lexical analysis phase:** Stream of characters (source program) acts as an input for the lexical analyzer, which produces the token stream as output (see Figure 1.8).

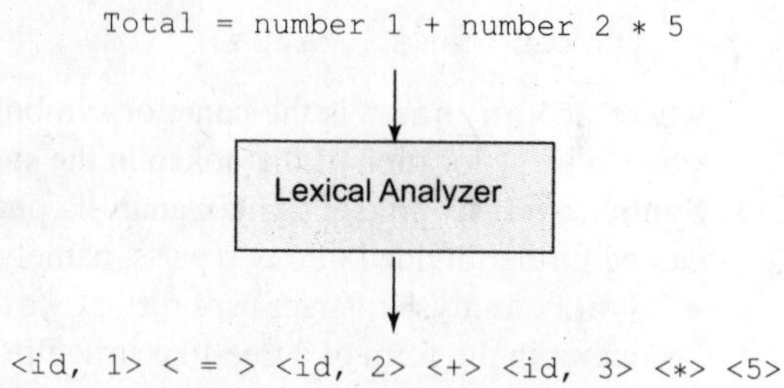

Figure 1.8 Lexical Analysis Phase

- **Syntax analysis phase:** The token stream acts as the input for the syntax analyzer. Output of the syntax analyzer is a parse tree (see Figure 1.9(a)) that acts as the input for the semantic analyzer; the output of the semantic analyzer is also a parse tree after type checking (see Figure 1.9(b)).

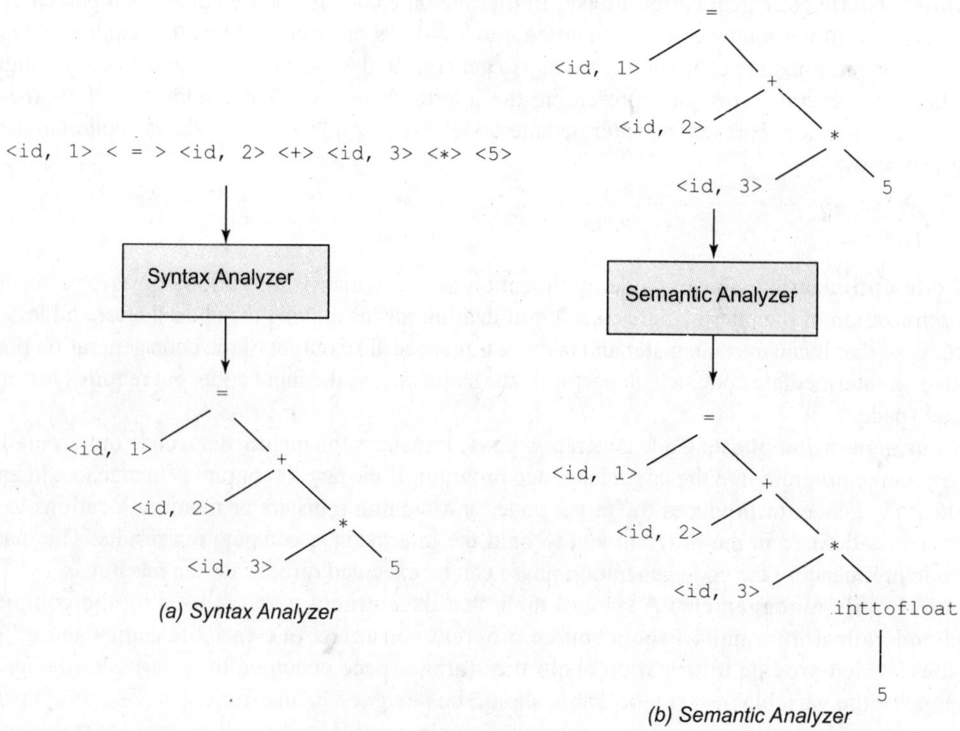

Figure 1.9 Syntax Analysis Phase

- **Intermediate code generation phase:** The parse tree acts as the input for the intermediate code generator, which produces an intermediate code as output (see Figure 1.10).
- **Code optimization phase:** The intermediate code of the source program acts as the input for the code optimizer. The output of the code optimizer is also an intermediate code (see Figure 1.11), that takes lesser space and lesser time to execute, and does the same task as the input intermediate code.
- **Code generation phase:** The optimized code acts as the input for the code generator. The output of the code generator is the machine language code (see Figure 1.12), known as the target program, which can be directly executed.

Note that the first operand in each instruction specifies a destination, and F in each instruction indicates that it deals with floating-point numbers.

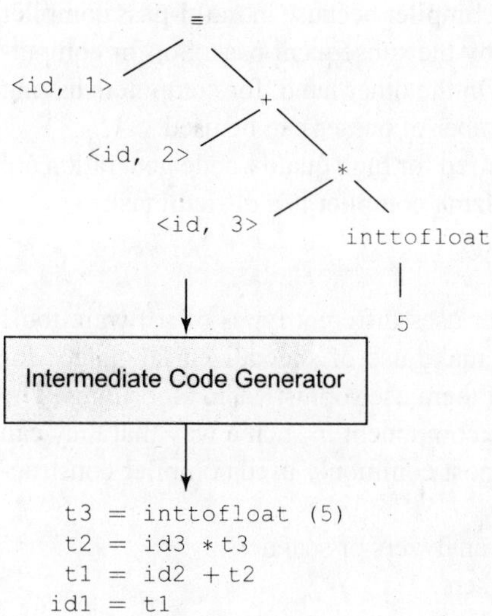

Figure 1.10 Intermediate Code Generation Phase

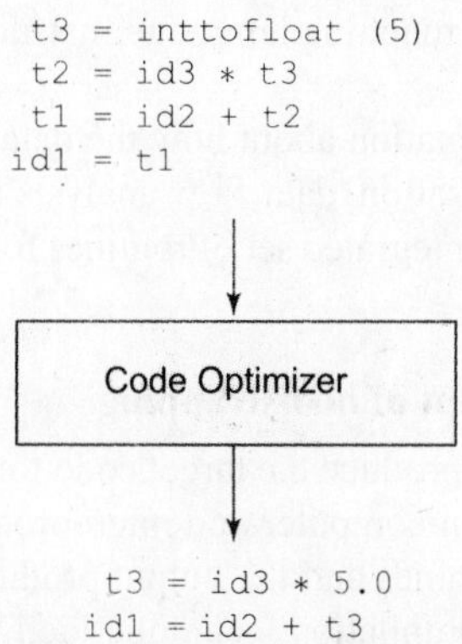

Figure 1.11 Code Optimization Phase

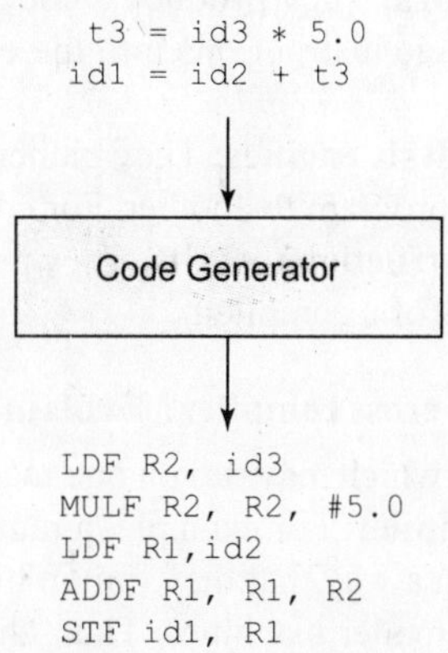

Figure 1.12 Code Generation Phase

8. What is a pass in the compilation process? Compare and contrast the features of a single-pass compiler with multi-pass compiler.

Ans: In an implementation of a compiler, the activities of one or more phases are combined into a single module known as a **pass**. A pass reads the input, either as a source program file or as the output of the previous pass, transforms the input and writes the output into an intermediate file. The intermediate file acts as either the input for the next pass or the final machine code.

When all the phases of a compiler are grouped together into a single pass, then that compiler is known as **single-pass compiler**. On the other hand, when different phases of a compiler are grouped together into two or more passes, then that compiler is known as **multi-pass compiler**.

A single-pass compiler is faster than the multi-pass compiler because in multi-pass compiler each pass reads and writes an intermediate file, which makes the compilation process time consuming. Hence, time required for compilation increases with the increase in the number of passes in a compiler.

A single-pass compiler takes more space than the multi-pass compiler because in multi-pass compiler the space used by the compiler during one pass can be reused by the subsequent pass. So, for computers having small memory, multi-pass compilers are preferred. On the other hand, for computers having large memory, single-pass compiler or compiler with fewer number of passes can be used.

In a single-pass compiler, the complicated optimizations required for high quality code generation are not possible. To count the exact number of passes for an optimizing compiler is a difficult task.

9. What are the various compiler construction tools?

Ans: For the construction of a compiler, the compiler writer uses different types of software tools that are known as **compiler construction tools**. These tools make use of specialized languages for specifying and implementing specific components, and most of them use sophisticated algorithms. The tools should hide the details of the algorithm used and produce component in such a way that they can be easily integrated into the rest of the compiler. Some of the most commonly used compiler construction tools are:

- **Scanner generators:** They automatically produce lexical analyzers or scanners.
- **Parser generators:** They produce syntax analyzers or parsers.
- **Syntax-directed translation engines:** They produce a collection of routines, which traverses the parse tree and generates the intermediate code.
- **Code generators:** They produce a code generator from a set of rules that translates the intermediate language instructions into the equivalent machine language instructions for the target machine.
- **Data-flow analysis engines:** They gather the information about how the data is transmitted from one part of the program to another. For code optimization, data-flow analysis is a key part.
- **Compiler-construction toolkits:** They provide an integrated set of routines for construction of the different phases of a compiler.

10. What is a cross compiler? Explain the concept of bootstrapping.

Ans: A compiler which may run on one machine and produce the target code for another machine is known as **cross compiler**. For example, a number of minicomputer and microprocessor compilers are implemented in such a way that they run on bigger machines and the output produced by them acts as an object code for smaller machines. Thus, the cross compilation technique facilitates platform independence. A cross compiler can be represented with the help of a T diagram as shown in Figure 1.13. It consists of three symbols S, T and I, where:

- S is the source language in which the source program is written,
- T is the target language in which the compiler produces its output or target program, and
- I is the implementation language in which compiler is written.

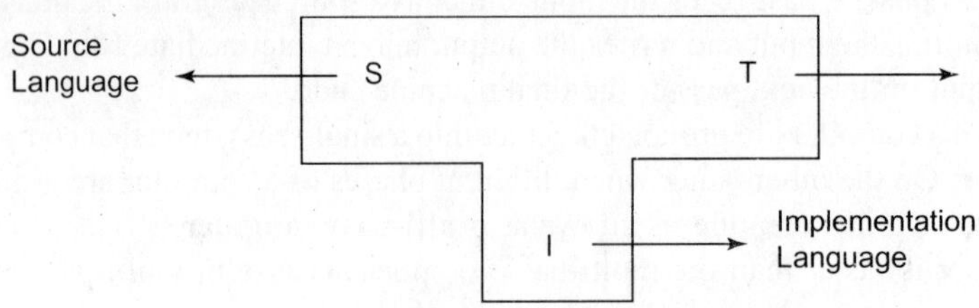

Figure 1.13 T Diagram Representation

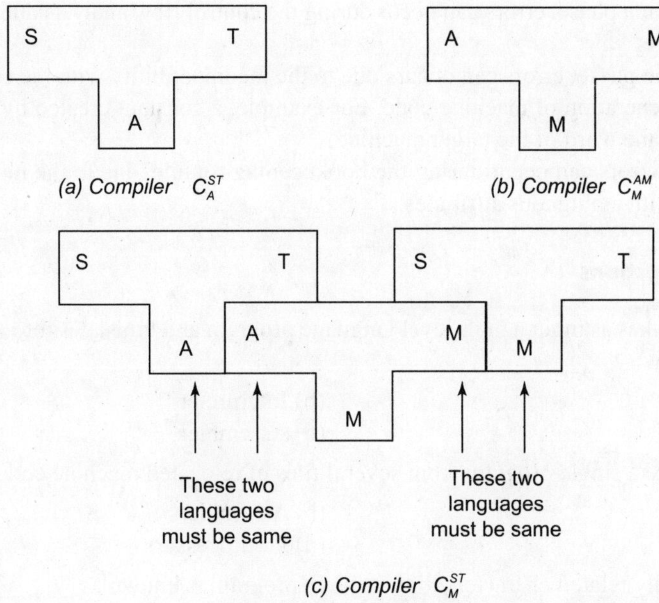

Figure 1.14 Bootstrapping

Bootstrapping: Bootstrapping is an important concept for building a new compiler. This concept uses a simple language to translate complicated programs which can further handle more complicated programs. The process of bootstrapping can be better understood with the help of an example given here.

Suppose we want to create a cross compiler for the new source language S that generates a target code in language T, and the implementation language of this compiler is A. We can represent this compiler as C_A^{ST} (see Figure 1.14(a)). Further, suppose we already have a compiler written for language A with both target and implementation language as M. This compiler can be represented as C_M^{AM} (see Figure 1.14(b)). Now, if we run C_A^{ST} with the help of C_M^{AM}, then we get a compiler C_M^{ST} (see Figure 1.14(c)). This compiler compiles a source program written in language S and generates the target code in T, which runs on machine M (that is, the implementation language for this compiler is M).

11. Explain error handling in compiler.

Ans: Error detection and reporting of errors are important functions of the compiler. Whenever an error is encountered during the compilation of the source program, an error handler is invoked. Error handler generates a suitable error reporting message regarding the error encountered. The error reporting message allows the programmer to find out the exact location of the error. Errors can be encountered at any phase of the compiler during compilation of the source program for several reasons such as:

- In lexical analysis phase, errors can occur due to misspelled tokens, unrecognized characters, etc. These errors are mostly the typing errors.
- In syntax analysis phase, errors can occur due to the syntactic violation of the language.
- In intermediate code generation phase, errors can occur due to incompatibility of operands type for an operator.

- In code optimization phase, errors can occur during the control flow analysis due to some unreachable statements.
- In code generation phase, errors can occurs due to the incompatibility with the computer architecture during the generation of machine code. For example, a constant created by compiler may be too large to fit in the word of the target machine.
- In symbol table, errors can occur during the bookkeeping routine, due to the multiple declaration of an identifier with ambiguous attributes.

Multiple-Choice Questions

1. A translator that takes as input a high-level language program and translates into machine language in one step is known as _________.

   (a) Compiler (b) Interpreter
   (c) Preprocessor (d) Assembler

2. _________ create a single program from several files of relocated machine code.

   (a) Loaders (b) Assemblers
   (c) Link editors (d) Preprocessors

3. A group of logically related characters in the source program is known as _________.

   (a) Token (b) Lexeme
   (c) Parse tree (d) Buffer

4. The _________ uses the parse tree and symbol table checking the semantic consistency of the source program.

   (a) Lexical analyzer (b) Intermediate code generator
   (c) Syntax translator (d) Semantic analyzer

5. The _________ phase converts an intermediate code into an optimized code that takes lesser space and lesser time to execute.

   (a) Code optimization (b) Syntax directed translation
   (c) Code generation (d) Intermediate code generation

6. _________ is invoked whenever any fault occurs in the compilation process of source program.

   (a) Syntax analyzer (b) Code generator
   (c) Error handler (d) Lexical analyzer

7. In compiler, the activities of one or more phases are combined into a single module known as a _________.

   (a) Phase (b) Pass
   (c) Token (d) Macro

8. For the construction of a compiler, the compiler writer uses different types of software tools that are known as _________.

   (a) Compiler writer tools (c) Programming tools
   (c) Compiler construction tools (d) None of these

9. A compiler that runs on one machine and produces the target code for another machine is known as ___________.
   (a) Cross compiler
   (b) Linker
   (c) Preprocessor
   (d) Assembler
10. If we run a compiler C_A^{ST} with the help of another compiler C_M^{AM}, then we get a new compiler that is ___________.
   (a) C_M^{SM}
   (b) C_A^{ST}
   (c) C_M^{ST}
   (d) C_M^{AM}

Answers

1. (a) 2. (c) 3. (b) 4. (d) 5. (a) 6. (c) 7. (b) 8. (c) 9. (a) 10. (c)

2

Lexical Analysis

1. What is the role of a lexical analyzer?

Ans: The lexical analysis is the first phase of a compiler where a lexical analyzer acts as an interface between the source program and the rest of the phases of compiler. It reads the input characters of the source program, groups them into lexemes, and produces a sequence of tokens for each lexeme. The tokens are then sent to the parser for syntax analysis. If lexical analyzer is placed as a separate pass in the compiler, it would require an intermediate file to place its output, from which the parser would then take its input. To eliminate the need for the intermediate file, the lexical analyzer and the syntactic analyzer (parser) are often grouped together into the same pass where the lexical analyzer operates either under the control of the parser or as a subroutine with the parser.

The parser requests the lexical analyzer for the next token, whenever it needs one. The lexical analyzer also interacts with the symbol table while passing tokens to the parser. Whenever a token is found, the lexical analyzer returns a representation for that token to the parser. If the token is a simple construct such as parentheses, comma, or a colon, then it returns an integer code. If the token is a more complex element such as an identifier or another token with a value, the value is also passed to the parser. The lexical analyzer provides this information by calling a bookkeeping routine which installs the actual value in the symbol table if it is not already there.

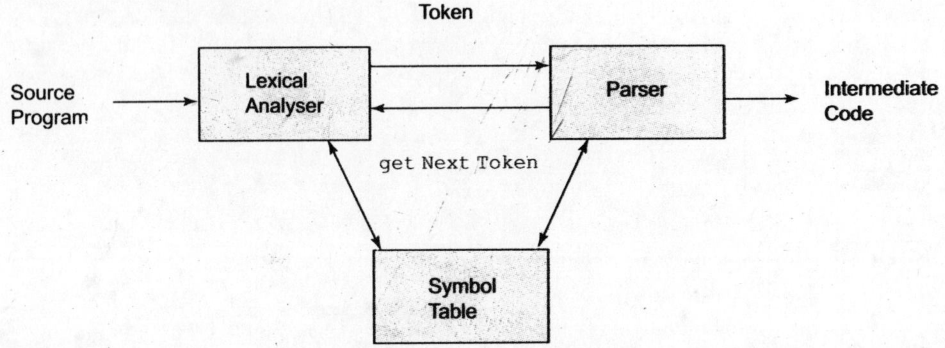

Figure 2.1 Role of the Lexical Analyzer

Besides generation of tokens, the lexical analyzer also performs certain other tasks such as:

- Stripping out comments and whitespace (tab, newline, blank, and other characters that are used to separate tokens in the input).
- Correlating error messages that are generated by the compiler during lexical analysis with the source program. For example, it can keep track of all newline characters so that it can associate an ambiguous statement line number with each error message.
- Performing the expansion of macros, in case macro preprocessors are used in the source program.

2. What do you understand by the terms tokens, patterns, and lexemes?

Ans: Tokens: The lexical analyzer separates the characters of the source language into groups that logically belong together, commonly known as **tokens**. A token consists of a token name and an optional attribute value. The **token name** is an abstract symbol that represents a kind of lexical unit and the optional attribute value is commonly referred to as **token value**. Each token represents a sequence of characters that can be treated as a single entity. Tokens can be identifiers, keywords, constants, operators, and punctuation symbols such as commas and parenthesis. In general, the tokens are broadly classified into two types:

- Specific strings such as if, else, comma, or a semicolon.
- Classes of strings such as identifiers, constants, or labels.

For example, consider an assignment statement in C

```
total = number1 + number2 * 5
```

After lexical analysis, the tokens generated are as follows:

```
<id,1> <=> <id,2> <+> <id,3> <*> <5>
```

Patterns: A rule that defines a set of input strings for which the same token is produced as output is known as **pattern**. Regular expressions play an important role for specifying patterns. If a keyword is considered as a token, the pattern is just the sequence of characters. But for identifiers and some other tokens, the pattern forms a complex structure.

Lexemes: A lexeme is a group of logically related characters in the source program that matches the pattern for a token. It is identified as an instance of that token by the lexical analyzer. For example, consider a C statement:

```
printf("Total = %d\n", total);
```

Here, `printf` is a keyword; parentheses, semicolon, and comma are punctuation symbols; `total` is a lexeme matching the pattern for token **id**; and `"Total = %d\n"` is a lexeme matching the pattern for token **literal**.

Some examples of tokens, patterns, and lexemes are given in Table 2.1.

Table 2.1 Examples of Tokens, Patterns and Lexemes

Token	Informal Description	Sample Lexeme
`while`	Characters `w, h, i, l, e`	`while`
`then`	Characters `t, h, e, n`	`then`
`comparison`	< or > or <= or >= or == or !=	<=, !=
`id`	Letter followed by letters and digits	`total, number1`
`number`	Any numeric constant	50, 3.12134, 0, 4.02e45
`literal`	Anything within double quotes(" ") except "	`"Total"`

3. What is the role of input buffering scheme in lexical analyzer?

Ans: The lexical analyzer scans the characters of source program one by one to find the tokens. Moreover, it needs to look ahead several characters beyond the next token to determine the next token itself. So, an input buffer is needed by the lexical analyzer to read its input. In a case of large source program, significant amount of time is required to process the characters during the compilation. To reduce the amount of overhead needed to process a single character from input character stream, specialized buffering techniques have been developed. An important technique that uses two input buffers that are reloaded alternately is shown in Figure 2.2.

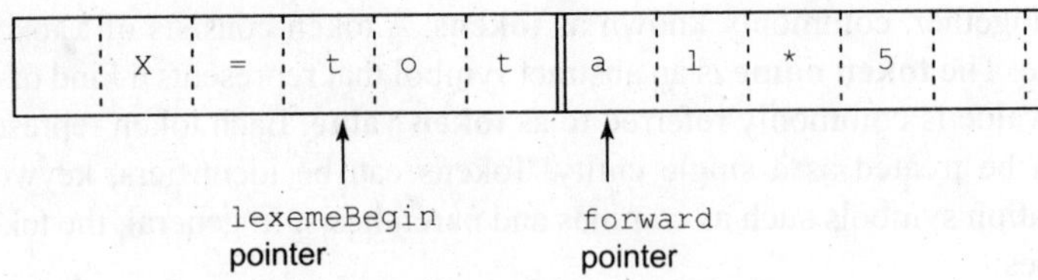

Figure 2.2 Input Buffer

Each buffer is of the same size `N`, where `N` is the size of a disk block, for example 1024 bytes. Thus, instead of one character, `N` characters can be read at a time. The pointers used in the input buffer for recognizing the lexeme are as follows:

- Pointer **`lexemeBegin`** points the beginning of the current lexeme being discovered.
- Pointer **`forward`** scans ahead until a pattern match is found for lexeme.

Initially, both pointers point to the first character of the next lexeme to be found. The `forward` pointer is scanned ahead until a match for a pattern is found. After the lexeme is processed, both the pointers are set to the character following that processed lexeme. For example, in Figure 2.2 the `lexemeBegin` pointer is at character `t` and `forward` pointer is at character `a`. The `forward` pointer is scanned until the lexeme `total` is found. Once it is found, both these pointers point to `*`, which is the next lexeme to be discovered.

4. What are strings and languages in lexical analysis? What are the operations performed on the languages?

Ans: Before defining the terms strings and languages, it is necessary to understand the term alphabet. An **alphabet** (or **character class**) denotes any finite set of symbols. Symbols include letters, digits, punctuation, etc. The ASCII, Unicode, and EBCDIC are the most important examples of alphabet. The set {0, 1} is the binary alphabet.

A **string** (also termed as **sentence** or **word**) is defined as a finite sequence of symbols drawn from an alphabet. The length of a string `s` is measured as the number of occurrences of symbols in `s` and is denoted by |`s`|. For example, the word 'orange' is a string of length six. The *empty string* ($\in$) is the string of length zero.

A **language** is any finite set of strings over some specific alphabet. This is an enormously broad definition. Simple sets such as ϕ, the empty set, or $\{\in\}$, the set containing only the empty string, are also the languages under this definition.

In lexical analysis, there are several important operations like union, concatenation, and closure that can be applied to languages. **Union** operation means taking all the strings of both the set of languages and creating a new set of language containing all the strings. The **concatenation** of languages is done

Table 2.2 Operations on Languages

Operation	Definition	Example: Let P = {A, B, . . . , Z, a, b, . . . , z} and Q = {0, 1, 2, . . . , 9}
Union of P and Q	P ∪ Q = {s | s is in P or s is in Q}	P ∪ Q is the set of letters and digits, with 62 strings of length one.
Concatenation of P and Q	PQ = {st | s is in P and t is in Q}	PQ is the set of 520 strings of length two, consisting one letter followed by one digit.
Kleene closure of P	$P^* = \cup_{i=0}^{\infty} P^i$	P^* is the set of all strings of letters, including ∈, the empty string. or $P(P \cup Q)^*$ is the set of all strings of letters and digits beginning with a letter.
Positive closure of Q	$Q^i = \cup_{i=0}^{\infty} Q^i$	Q^+ is the set of all strings of one or more digits.

by concatenating a string from the first language and a string from the second language forming the new strings, in all possible ways. The (Kleene) closure of a language P, denoted by P^*, is the set of strings achieved by concatenating P zero or more times. P^0, 'the concatenation of P zero times,' is defined to be {∈}. The positive closure, denoted by P^+, is same as the Kleene closure, but without the term P^0, precisely, P^+ is $P^{i-1}P$. ∈ will not be in P^+, unless it is in P itself. These operations are listed in Table 2.2.

5. Define the following terms in context of a string: prefix, suffix, substring, and subsequence.

Ans: Prefix: If zero or more symbols are removed from the end of any string s, a new string is obtained known as a prefix of string s. For example, app, apple, and ∈ are prefixes of apple.

Suffix: If zero or more symbols are removed from the beginning of any string s, a new string is obtained known as suffix of string s. For example, ple, apple, and ∈ are suffixes of apple.

Substring: If we delete any prefix and any suffix from a string s, we obtain a new string known as substring of s. For example, pp, apple, and ∈ are substrings of apple.

Subsequence: If we delete zero or more not necessarily consecutive positions of a string s, a new string is formed known as subsequence of s. For example, ale is a subsequence of apple.

6. What do you mean by a regular expression? Write a regular expression over alphabet Σ = (x, y, z) that represents all strings of length three.

Ans: A **regular expression** is a compact notation that is used to represent the patterns corresponding to a token. It is used to describe all the languages that can be built by applying union, concatenation, and closure operations to the symbols of some alphabet. The regular expression represents pattern to define the language which includes a set of strings. The strings are considered to be in the said language if they match the pattern; otherwise, they are not in the said language.

For example, consider the identifiers in a programming language, where an identifier may consist of a letter or more followed by any number of digits or an underscore (_). Thus, the language for C identifiers can be described as:

```
letter_(letter_|digit)*
```

Here, the vertical bar indicates union and the star indicates zero or more instances. The parentheses are used to group subexpressions.

There exist some primitive regular expressions which are of universal type, over some alphabet Σ, which are defined as follows:

- x (for each $x \in \Sigma$), the primitive regular expression x defines the language {x}, that is, the only string is 'x' in this particular language which is of length one.
- λ (empty string), the primitive regular expression λ defines the language $\{\lambda\}$, that is, the only string is the empty string in this particular language. The language denoted by λ is of universal type.
- ϕ (indicates no string at all), the primitive regular expression ϕ denotes the language { }, that is, no string at all in this particular language. The language denoted by ϕ is also of universal type.

Thus, it must be noted that if $|\Sigma|$ = number of symbols present in it = n, then there are n + 2 primitive regular expressions defined over it.

Construction of Regular Expression

Given Σ = {x, y, z}.

And, we have to construct a regular expression that represents all strings of length three.

For this, let us choose three arbitrary symbols a_1, a_2, a_3.

Thus, the regular expression will be:

$a_1a_2a_3$,

where

a_1 = either 'x' or 'y' or 'z'
a_2 = either 'x' or 'y' or 'z'
a_3 = either 'x' or 'y' or 'z'

7. List the rules for constructing regular expressions. Write some properties to compose additional regular expressions. What is a regular definition? Give a suitable example.

Ans: The rules for constructing regular expressions over some alphabet Σ are divided into two major classifications which are as follows:

(i) Basic rules (ii) Induction rules

Basic rules: There are two rules that form the basis:

1. $\in$ is a regular expression, and $L(\in)$ is $\{\in\}$, that is, its language contains only an empty string.
2. If a is a symbol in Σ, then a is a regular expression, and $L(a) = \{a\}$, which implies the language with one string, of length one, with a in its one position.

Induction rules: There are four induction rules that built larger regular expressions recursively from smaller regular expressions. Suppose R and S are regular expressions with languages L(R) and L(S), respectively.

1. (R)(S) is a regular expression representing the language L(R).L(S).
2. (R) | (S) is a regular expression representing the language $L(R) \cup L(S)$.
3. $(R)^*$ is a regular expression representing the language $(L(R))^*$.
4. (R) is a regular expression representing L(R). This rule states that additional pairs of parentheses can be added around expressions without modifying the language.

Properties of Regular Expression: To compose additional regular expressions, the following properties are to be considered, a finite number of times:

1. If a_1 is a regular expression, then (a_1) is also a regular expression.
2. If a_1 is a regular expression, then $a_1{}^*$ is also a regular expression.
3. If a_1 and a_2 are two regular expressions, then a_1a_2 is also a regular expression.
4. If a_1 and a_2 are two regular expressions, then $a_1 + a_2$ is also a regular expression.

Regular Definition: If Σ = alphabet set, then a regular definition is a sequence of definitions of the form:

$$D_1 \rightarrow R_1$$
$$D_2 \rightarrow R_2$$
$$\ldots$$
$$D_n \rightarrow R_n$$

where

- ❑ D_i is a new symbol, not in Σ and not the same as any of the other D's.
- ❑ R_i is a regular expression over the alphabet $\Sigma \cup \{D_1, D_2, \ldots, D_{i-1}\}$.

For example, let us consider the C identifiers that are strings of letters, digits, and underscores. Here, we give a regular definition for the language of C identifiers.

```
letter_  → A| B |...| Z | a | b |...| z | _
digit → 0 | 1 |...| 9
id → letter_(letter_|digit)*
```

8. What is a transition diagram? Draw a transition diagram to identify the keywords `IF`, `THEN`, `ELSE`, `DO`, `WHILE`, `BEGIN`, `END`.

Ans: While constructing a lexical analyzer, we represent patterns in the form of flowcharts, called transition diagrams. A **transition diagram** consists of a set of nodes and edges that connect one state to another. A node (or a circle) in a transition diagram represents a state and each edge (or an arrow) represents the transition from one state to another. Each edge is labeled with one or more symbols.

A **state** is basically a condition that could occur while scanning the input to find out a lexeme that matches one of the several patterns. We can also think of a state as summarizing all we need to know what characters have been seen between the `lexemeBegin` pointer and the `forward` pointer. Suppose, currently we are at state q, and the next input symbol is a, then we look for an edge e coming out of the current state q that is having the label a. If such an edge is found, then we move ahead the `forward` pointer and enter the state of the transition diagram to which this edge is connected.

Among all the states, one state, say q_0, is termed as initial or start state. The transition diagram always begins in the start state before any input symbols have been read. One or more states are said to be final or accepting states and are represented by double circles. We may also attach actions to the final states to indicate that a token and an attribute value are being returned to the parser. In some cases, it is also necessary to move the `forward` pointer backward by certain number of positions, then we can place that many number of *'s near the final state. For example, if we want to retract the pointer by one position, then we can place a single *, for two positions, ** can be placed, and so on.

The transition diagram to identify the keywords BEGIN, END, IF, THEN, ELSE, DO, and WHILE is shown in Figure 2.3.

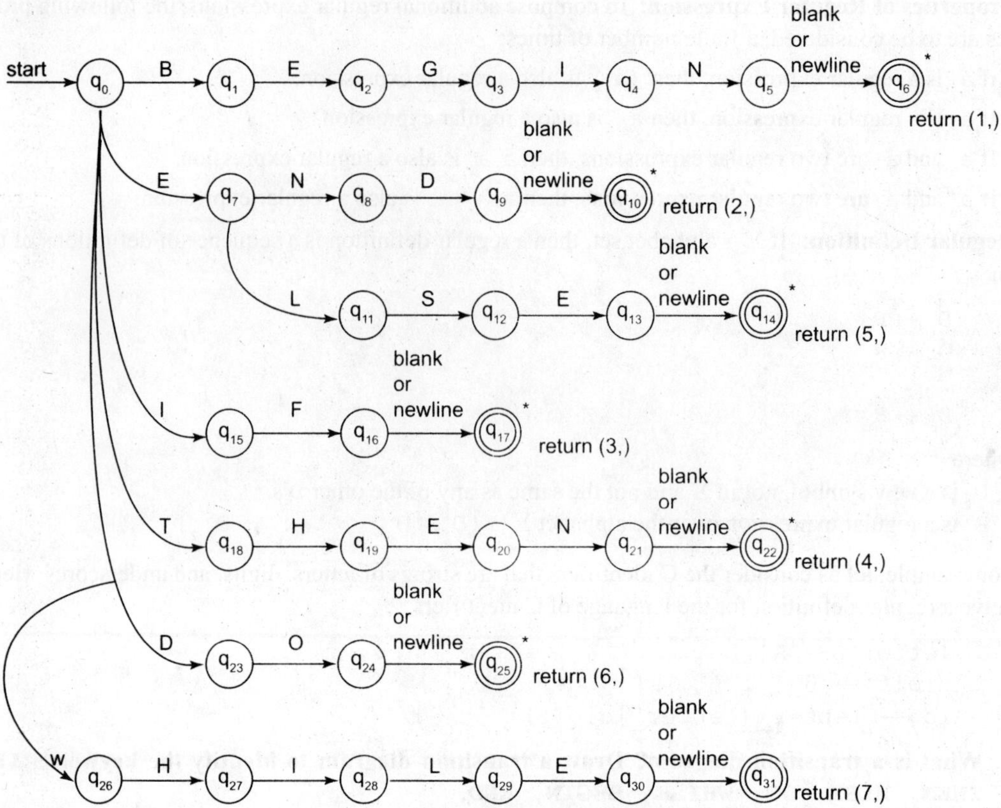

Figure 2.3 Transition Diagram to Identify Keywords

9. Draw the transition diagram for identifiers, constants, and relational operators (relops).

Ans: Transition diagram for identifiers is shown in Figure 2.4.

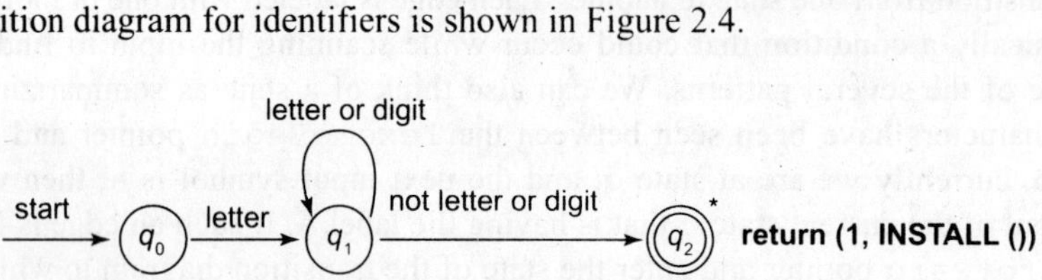

Figure 2.4 Transition Diagram for Identifiers

The transition diagram for constants is shown in Figure 2.5.

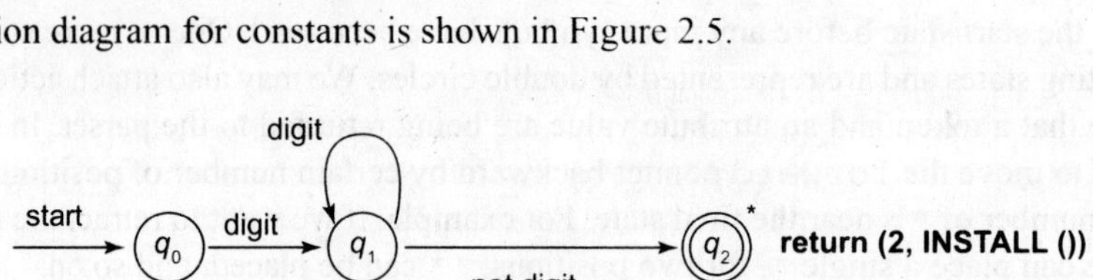

Figure 2.5 Transition Diagram for Constants

The transition diagram for relational operators (relops) is shown in Figure 2.6.

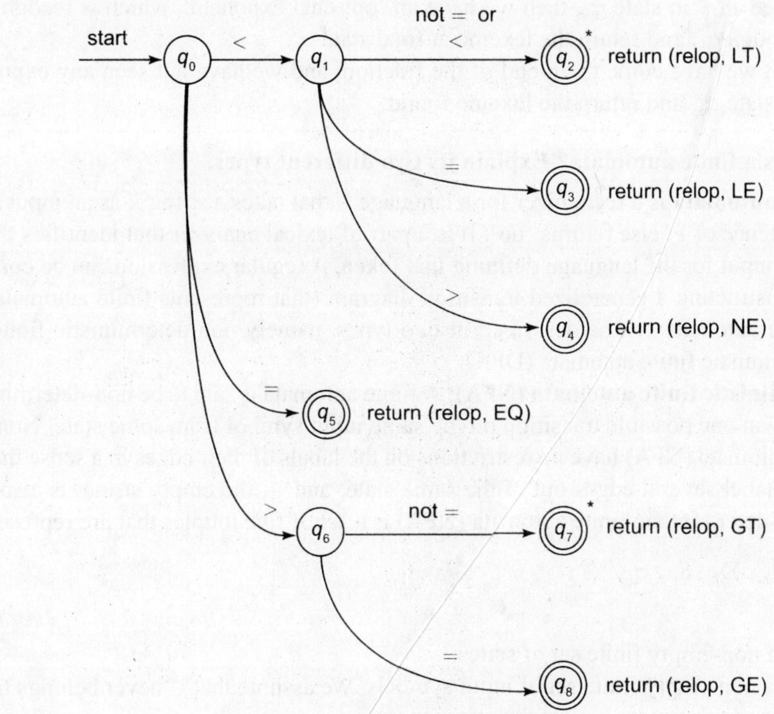

Figure 2.6 Transition Diagram for Relops

10. Draw a transition diagram for unsigned numbers.

Ans: The transition diagram for unsigned numbers is shown in Figure 2.7.

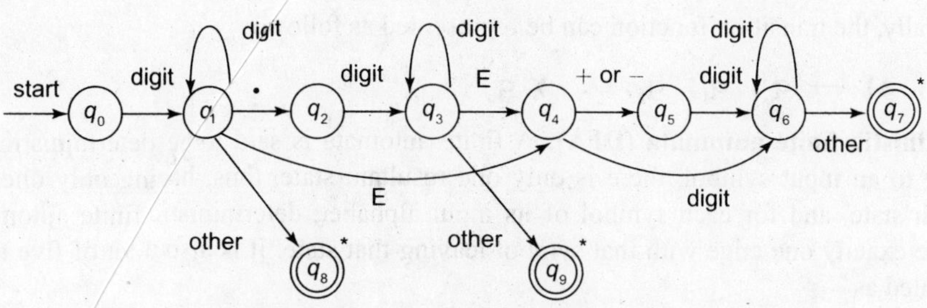

Figure 2.7 Transition Diagram for Unsigned Numbers

- In the transition diagram for unsigned numbers, we begin with the start state q_0, if we see a digit, we move to state q_1. In that state, we can read any number of additional digits.
- In case we see anything except a digit, dot, or E from state q_1, it implies that we have seen an integer number, for example 789. In such case, we enter the state q_8, where we return token number and a pointer to a table of constants where lexeme is entered.

- If we see a dot from state q_1, then we have an 'optional fraction,' and we enter the state q_2. Now if we look for one or more additional digits, we move to the state q_3 for this purpose.
- In case we see an E in state q_3, then we have an 'optional exponent,' which is recognized by the states q_4 through q_7, and return the lexeme at final state q_7.
- In state q_3, if we have come to an end of the fraction, and we have not seen any exponent E, we move to the state q_9, and return the lexeme found.

11. What is a finite automata? Explain its two different types.

Ans: A **finite automata** is a recognizer for a language P that takes a string x as an input and returns 'yes' if x is a sentence of P, else returns 'no'. It is a part of lexical analyzer that identifies the presence of a token on the input for the language defining that token. A regular expression can be converted to a recognizer by constructing a generalized transition diagram (that represents finite automata) from the expression. Finite automata can be described in two types, namely non-deterministic finite automata (NFA) and deterministic finite automata (DFA).

- **Non-deterministic finite automata (NFA):** A finite automata is said to be non-deterministic, if we have more than one possible transition on the same input symbol from some state. Non-deterministic finite automata (NFA) have no restrictions on the labels of their edges in a sense that the same symbol can label several edges out of the same state, and $\in$, the empty string, is also a possible label. A non-deterministic finite automata (NFA) is a set of five touples that are represented as:

  $$M = (Q, \Sigma, \delta, q_0, F)$$

  where

  Q is a non-empty finite set of states.
  Σ is a non-empty finite set of input symbols. We assume that $\in$ never belongs to Σ.
  q_0 is a starting state, one of the states in Q.
  F is a subset of Q containing final (or accepting) states.
  δ is a transition function, which takes two arguments, a state and an input symbol from Σ $\cup\{\in\}$, and returns a set of next states. δ is represented as:

  $$\delta : Q * (\Sigma \cup \{\in\}) \rightarrow 2^Q$$

  Graphically, the transition function can be represented as follows:

  $$\delta(q, a) \rightarrow q_0, q_1, q_2, \ldots, q_n$$

- **Deterministic finite automata (DFA):** A finite automata is said to be deterministic, if corresponding to an input symbol, there is only one resultant state, thus, having only one transition. For each state, and for each symbol of its input alphabet, deterministic finite automata (DFA) can have exactly one edge with that symbol leaving that state. It is also a set of five touples and represented as

  $$M = (Q, \Sigma, \delta, q_0, F)$$

  where

  Q is a non-empty finite set of states.
  Σ is a non-empty finite set of input symbols.
  q_0 is an initial state of DFA and member of Q.
  F is a subset of Q containing final states.

δ is transition function, which takes two arguments, a state and an input symbol, and returns a single state 'represented by `Q * Σ→ Q`'. Let `q` is the state and `a` be the input symbol passed to the transition function, then `δ(q, a) = q'`, where `q'` is the output function, which may be same as `q`.

Graphically, the transition function can be represented as follows:

δ (q, a) → q'

DFA is a special case of an NFA where

- There are no moves on input ∈ and
- For each state `q` and input symbol `a`, there is exactly one edge out of `q` labeled `a`.

12. What do you mean by NFA with Є-transition?

Ans: NFA with ∈-transition is defined as a modified finite automata that permits transition without input symbols, along with zero, one or more transitions on input symbols. Let us take an example, where we have to design an NFA with ∈-transition for the following accepting language:

`L = {ab U aab*}`

To solve this problem, first we divide the language as follows:

$L = L_1 \cup L_2$, where $L_1 = ab$ and $L_2 = aab^*$

Now, we construct NFA for L_1.

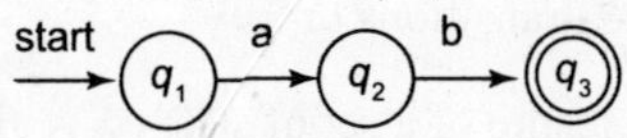

Now, we construct NFA for L_2.

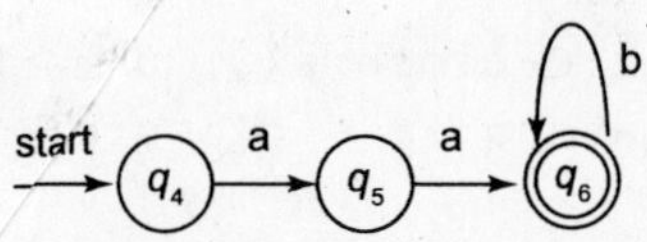

Finally, we combine the transition diagram of L_1 and L_2, to construct the NFA with ∈-transition for given input language as shown in Figure 2.8. In this NFA, we use ∈-transitions to reach at states q_1 and q_2.

13. Explain the working of Є-closure, with a suitable example.

Ans: NFA with ∈-transition accepts a string `w` in Σ^*, if there exists at least one path which corresponds to `w` initializing from start state and ending at final state. If the path contains ∈-moves, then we define a function `∈-closure(q)`, where `q` is the state of automata. The ∈-closure function is defined as follows:

`∈-closure(q)` = set of all those states of automata which can be reached from `q` on a path labeled by ∈, that is without consuming any input symbol. For example, consider the following NFA:

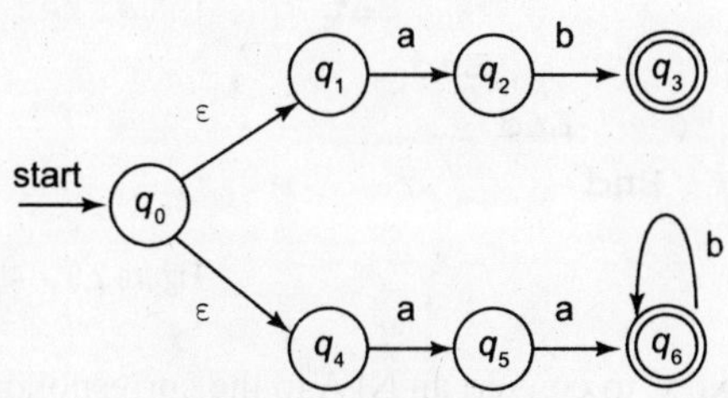

Figure 2.8 NFA with ∈-Transition

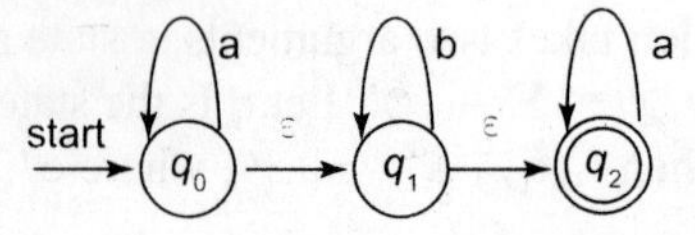

In this NFA,

$\in$-closure(q_0) = {q_0, q_1, q_2}
$\in$-closure(q_1) = {q_1, q_2}
$\in$-closure(q_2) = {q_2}

14. Write an algorithm to convert a given NFA into an equivalent DFA.

Or

Give the algorithm for subset construction and computation of ∈-closure.

Ans: The basic idea behind constructing a DFA from NFA is to merge two or more states of DFA into one. To convert a given NFA into an equivalent DFA, we note that a set of states in an NFA corresponds to a state in the DFA. All the NFA states are reachable from at least one state of the same set using ∈-transition only, without considering any further input. Moreover, from this set of states which are based on some input symbol we can reach another set of states. In the DFA, we take these sets as unique states. We define two sets that are as follows:

- **∈-closure(q):** In an NFA, ∈-closure of a state q defined to be the set of states (including q) that are reachable from q using ∈-transitions only.
- **∈-closure(Q):** ∈-closure of a set of states Q of an NFA is defined to be the set of states reachable from any state in Q using ∈-transitions only.

The algorithm for computing ∈-closure of a set of states Q is given in Figure 2.9.

```
∈-closure(Q) = Q
Set all the states of ∈-closure(Q) unmarked
For each unmarked state q in ∈-closure(Q) do
Begin
    Mark q
    For each state q' having an edge from q. to q' labeled ∈ do
    Begin
        If q' is not in ∈-closure(Q) then
        Begin
            add q' to ∈-closure(Q)
            Set q' unmarked
        End
    End
End
```

Figure 2.9 Algorithm for Computing ∈-Closure

Now, to convert an NFA to the corresponding DFA, we consider the algorithm shown in Figure 2.10.
Input: An NFA with set of states Q, start state q_0, set of final states F
Output: Corresponding DFA with start state d_0, set of states Q_D, set of final states F_D

```
Begin
    d0 = ∈-closure(q0)
    QD = {d0}
    If d0 contains a state from F then FD = {d0} else FD = ϕ
    Set d0 unmarked
    While there are unmarked states in QD do
    Begin
        Let d be such a state
        For each input symbol x do
        Begin
            Let S be the set of states in Q having transitions on x from
            any state of the NFA corresponding to the DFA state d
            d' = ∈-closure(S)
            If d' is already present in QD then
                add the transition d → d' labeled x
            else
            Begin
            QD = QD ∪ {d'}
            add the transition d → d' labeled x
            Set d' unmarked
            If d' contains a state of F then FD = FD ∪ {d'}
            End
        End
    End
End
```

Figure 2.10 Algorithm to Convert NFA to DFA

15. Give Thompson's construction algorithm. Explain the process of constructing an NFA from a regular expression.

Ans: To construct an NFA from a regular expression, we present a technique that can be used as a recognizer for the tokens corresponding to a regular expression. In this technique, a regular expression is first broken into simpler subexpressions, then the corresponding NFA are constructed and finally, these small NFAs are combined with the help of regular expression operations. This construction is known as **Thompson's construction**.

Thompson's construction algorithm: The brief description of Thompson's construction algorithm is as follows:

Step 1: Find the alphabet set Σ from the given regular expression. For example, for the regular expression `a (a | b)` * ab, Σ = `{a,b}`. Now, determine all primitive regular expressions.

Step 2: Construct equivalent NFAs for all primitive regular expressions. For example, an equivalent NFA for the primitive regular expression 'a' is shown below:

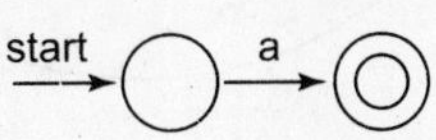

Step 3: Apply the rules for union, concatenation, grouping, and (Kleene)* to get the equivalent NFA of the given regular expression.

While constructing an NFA from a regular expression using Thompson's construction, these rules are followed:

- For $\in$ or any alphabet symbol x in the alphabet set Σ, the NFA consists of two states—a start state and a final state. The transition is labeled by $\in$ or x as shown below:

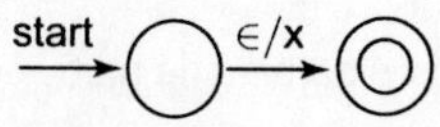

- If we are given NFAs of two regular expressions r_1 and r_2 as $N(r_1)$ and $N(r_2)$, then we can construct a composite NFA for the regular expression $(r_1 | r_2)$ as follows:
  - Add new initial state (q_0) and final state q_f.
  - Introduce $\in$-transitions from q_0 to the start state of $N(r_1)$ and $N(r_2)$. Similarly, introduce $\in$-transitions from final states of $N(r_1)$ and $N(r_2)$ to the new final state q_f (see Figure 2.11).

    Note that the final states of $N(r_1)$ and $N(r_2)$ are no longer be the final states in the composite NFA $N(r_1 | r_2)$.

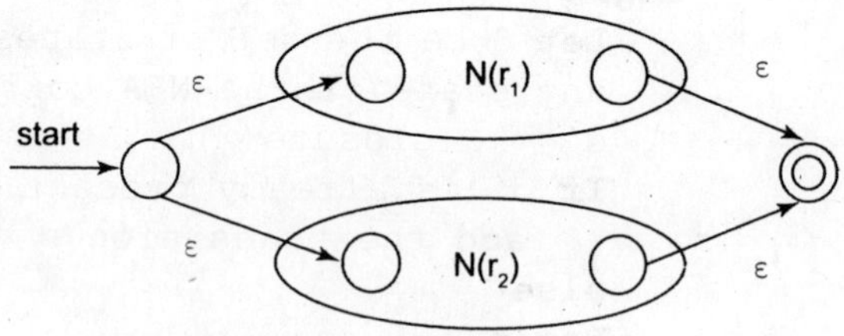

Figure 2.11 NFA for $r_1 | r_2$

- The NFA $N(r_1r_2)$ for the regular expression r_1r_2 can be constructed by merging the final state of $N(r_1)$ with the start state of $N(r_2)$. The start state of $N(r_1)$ will become the start state of new NFA and the final state of $N(r_2)$ will become the final state of new NFA as shown in Figure 2.12.

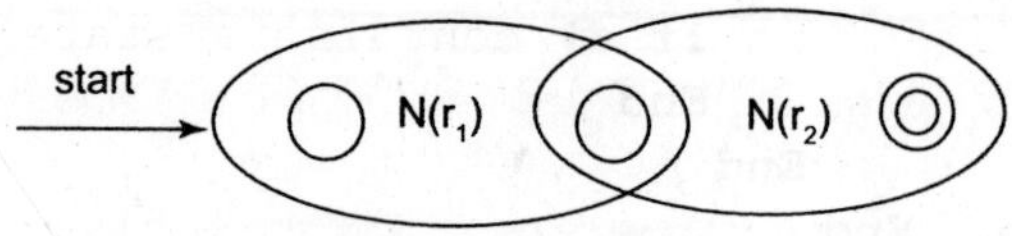

Figure 2.12 NFA for r_1r_2

- If we are given NFA $N(r^*)$, we construct a regular expression r^* from the NFA $N(r)$ of r as follows:
  - Add new start state (q_0) and final state (q_f).
  - Introduce $\in$-transitions from q_0 to the start state of $N(r)$, from the final state of $N(r)$ to q_f, from the final state of $N(r)$ back to the start state of $N(r)$ that corresponds to repeated occurrence of r, and from q_0 to q_f corresponding to the zero-occurrence of r (see Figure 2.13).
- If $N(r)$ be the NFA for a regular expression r, it is also the NFA for the parenthesized expression (r).

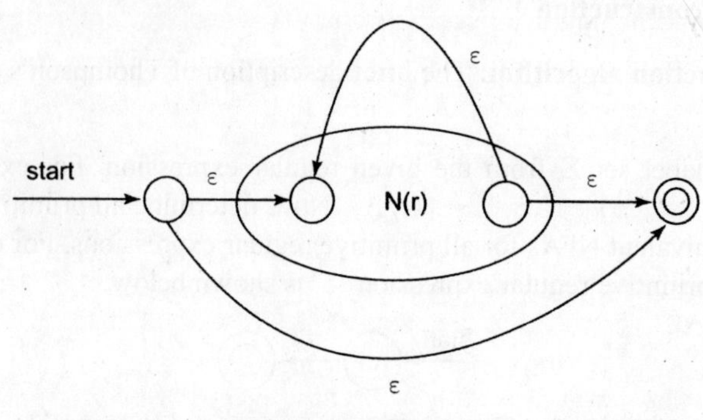

Figure 2.13 NFA for r*

16. Explain the functions `nullable(n)`, `firstpos(n)`, `lastpos(n)`, and `followpos(p)` and describe the rules to compute them.

Ans: To convert a regular expression into DFA, we construct syntax tree of the regular expression and then compute the four functions as follows:

`nullable(n)`: This function is true for a syntax-tree node n if and only if its subexpression contains ∈ in its language. In other words, the subexpression can be *made null* or the empty string, even if it can represent other strings as well. The rules to compute `nullable(n)` for any node n are given as follows:

- For a leaf labeled ∈, `nullable` is true.
- For a leaf with position i, `nullable` is false because they correspond to non-∈ operands.
- For an or-node n = $c_1 | c_2$, `nullable` will be true only if either of its child is nullable.
- For a cat-node n = c_1c_2, `nullable` will be true only if both the children are nullable.
- For a star-node n = c_1*, `nullable` is always true.

`firstpos(n)`: It is the set of positions in the subtree rooted at n corresponding to the first symbol of at least one string in the language of the subexpression rooted at n. The rules to compute `firstpos(n)` for any node n are as follows:

- For a leaf labeled ∈, `firstpos(n)` will be ϕ.
- For a leaf with position i, `firstpos(n)` will be i itself.
- For an or-node n = $c_1 | c_2$, we take the union of the `firstpos` of left child and right child.
- For a cat-node n = c_1c_2, if the left child c_1 is `nullable`, then we take the union of `firstpos` of the left child c_1 as well as the right child c_2, otherwise only `firstpos` of the left child c_1 is possible.
- For star-node n = c_1*, we take the value of `firstpos` of the left child c_1.

`lastpos(n)`: It is the set of positions in the subtree rooted at n corresponding to the last symbol of at least one string in the language of the subexpression rooted at n. The rules to compute `lastpos` are the same as that of `firstpos`, except the rule for the cat-node, where the roles of its children are interchanged. That is, for a cat-node n = c_1c_2, we consider whether the right child c_2 is `nullable`. If yes, then we take the union of `lastpos(`c_1`)` and `lastpos(`c_2`)`, otherwise only `lastpos(`c_2`)` is possible.

`followpos(p)`: It is set of positions q, for a position p, in the syntax tree such that there exist some string s = $x_1x_2 \ldots x_n$ in `L((r)#)` such that for some i, there is a way to explain the membership of s in `L((r)#)` by matching x_i to position p of the syntax tree and x_{i+1} to position q. To compute `followpos`, there are only two ways given as follows:

- If n = c_1c_2, then for every position i in `lastpos(`c_1`)`, `followpos(i)` will be all positions in `firstpos(`c_2`)`.
- If n is a star-node, and i is a position in `lastpos(n)`, then `followspos(i)` will be all positions in `firstpos(n)`.

To understand how to compute these functions, consider the syntax tree for the expression `(x|y) * xyy#` shown in Figure 2.14. The numeric value associated with each leaf node indicates the position of the leaf and also the position of its symbol.

In this syntax tree, only the star-node is nullable because every star-node is nullable. All the leaf nodes correspond to non-∈ operands; thus, none of them is nullable. The or-node is also not nullable because neither of its child nodes is nullable. Finally, the cat-nodes also have non-nullable child nodes, and hence none of them is nullable. The `firstpos` and `lastpos` of all the nodes are shown in Figure 2.15.

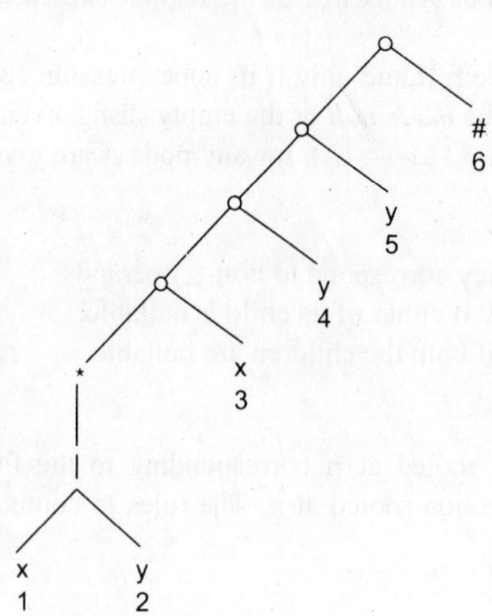

Figure 2.14 Syntax Tree for (x | y) * xyy#

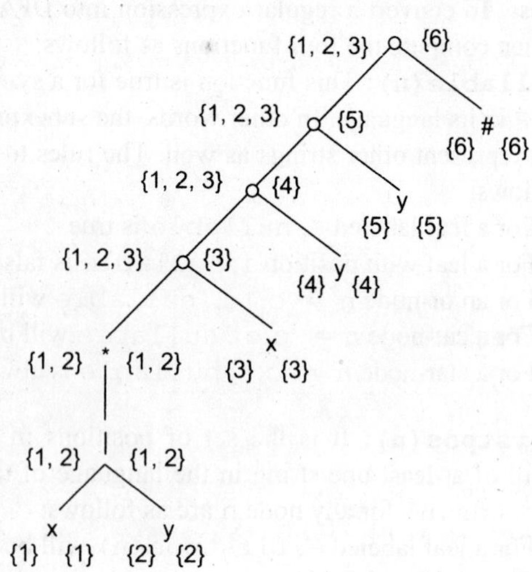

Figure 2.15 `Firstpos` and `Lastpos` for the Nodes

The `followpos` of all the leaf nodes is given in Table 2.3.

Table 2.3 `followpos` for the Nodes

Value of n	followpos(n)
1	{1,2,3}
2	{1,2,3}
3	{4}
4	{5}
5	{6}
6	φ

17. Describe the process of constructing a DFA directly from a regular expression.

Ans: The process for constructing a DFA directly from a regular expression consists of the following steps:

- From the augmented regular expression `(r)#`, construct a syntax tree `T` rooted at node n_0.
- For syntax tree `T`, compute `nullable`, `firstpos`, `lastpos`, and `followpos`.
- Construct `Dstates`, the set of states of DFA `D`, and `Dtran`, the transition function for `D`, by using the algorithm given in Figure 2.16.

The states of `D` are sets of position in `T`. Initially, all the states are unmarked, and a state becomes marked just before its out-transitions. `firstpos`(n_0) is set as the start state of `D`, and the states containing the position for the endmarker symbol # are considered as the accepting states.

```
Initialize Dstates with only the unmarked state firstpos(n0)
For each unmarked state S in Dstates do
Begin
   Mark S
   For each input symbol x do
   Begin
      Let U be the union of followpos(p) for all p in S that cor
      respond to x
      if (U is not in Dstates) then
      add U as an unmarked state to Dstates
      Dtran[S,x] = U
   End
End
```

Figure 2.16 Algorithm for Constructing `Dstates` and `Dtran`

18. Explain lexical analyzer generator (LEX) and its structure.

Or

What is a lex compiler? Write its specification.

Ans: A **lex compiler** or simply **lex** is a tool for automatically generating a lexical analyzer for a language. It is an integrated utility of the UNIX operating system. The input notation for the lex is referred to as the **lex language**. The process of constructing a lex analyzer with the lex compiler is shown in Figure 2.17.

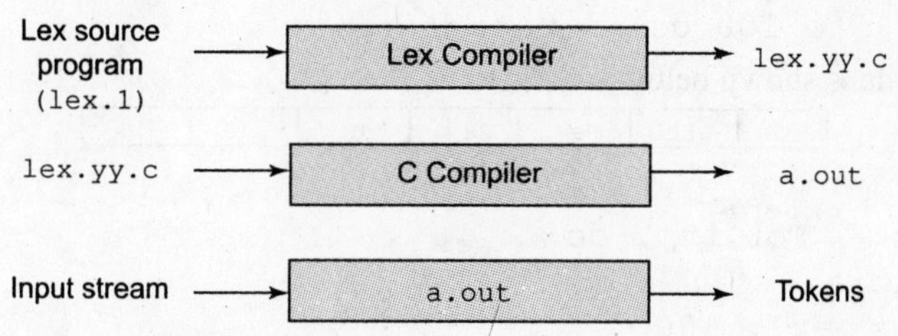

Figure 2.17 Constructing a Lexical Analyzer with Lex

The lex source program, `lex.1`, is passed through lex compiler to produce the C program file `lex.yy.c`. The file `lex.1` basically contains a set of regular expressions along with the routines for each regular expression. The routines contain a set of instructions that need to be executed whenever a token specified in the regular expression is recognized. The file `lex.yy.c` is then compiled using a C compiler to produce the lexical analyzer `a.out`. This lexical analyzer can now take a stream of input characters and produce a stream of tokens.

The lexical analyzer `a.out` is basically a function that is used as a subroutine of the parser. It returns an integer code for one of the possible token names. The attribute value for the token is stored in a global variable `yylval`. This variable is shared by both lexical analyzer and parser. This enables to return both the name and the attribute value of a token.

Lex Specifications or Structure

A lex program consists of the following form:

```
declarations
%%
translation rules
%%
auxiliary routines
```

The *declarations* section includes variable declarations, constant declarations, and regular definitions. The regular definitions can be used as parts in the *translation rules* section. This section contains the patterns and the associated action. The translation rules each have the following form:

```
Pattern {Action}
```

Here, each *pattern* is a regular expression, which may use the regular definitions of the declaration section. Each *action* specifies a set of statements to be executed whenever rule r_i matches the current input sequence.

The third section (*auxiliary routines*) holds the additional functions that may be used to write the `action` part. These functions can also be compiled separately and loaded with the lexical analyzer.

19. What are the proper recovery actions in lexical analysis?

Ans: The possible error recovery actions in lexical analysis phase are as follows:

- Deleting an extra character.
- Inserting a missing character.
- Replacing an incorrect character by a correct character.
- Transposing two adjacent characters.

20. Find the tokens for the given code:

```
For I = 1 to 100 do
```

Ans: The given code is shown below:

For	I	=	1	to	100	do

Tokens are:

```
Keywords → For, to, do
Identifiers → I
Constants → 1, 100
Operators → =
```

21. Construct a symbol table and find the tokens for the given code:

```
IF (i = 20) Then GOTO 100
```

Ans: The given code is shown below:

If	(	i	=	20	)	Then	GOTO	100

Tokens are:

```
Keywords → If, Then, GOTO
Identifiers → i
Constants → 20, 100
Operators → (, =, )
```

The symbol table corresponding to the given code is as follows:

	.
	.
231	constant, integer, value = 20
	.
	.
642	label, value = 100
	.
	.
782	identifier, integer, value = i

After finding the required tokens and storing them into the symbol table, code is rewritten as follows:

```
If([identifier, 782] = [constant, 231]) Then GOTO [label, 642]
```

22. Design a Finite Automata that accepts set of strings such that every string ends with 00, over alphabets {0,1}.

Ans: Here, we have to construct a finite automata that will accept all strings like {00, 01100, 110100, . . .}. The finite automata for the given problem is given below:

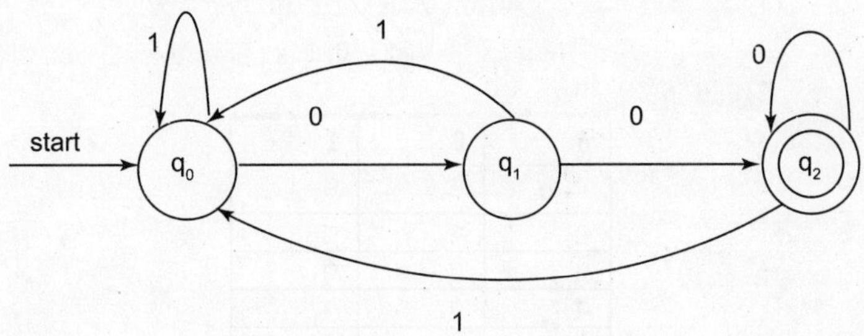

Hence, finite automata $M = \{Q, \Sigma, \delta, q_0, F\}$ will be

$Q = \{q_0, q_1, q_2\}$
$\Sigma = \{0, 1\}$
$q_0 = \{q_0\}$
$F = \{q_2\}$

The transition function δ is shown with the help of the following transition table:

δ	0	1
$\rightarrow q_0$	q_1	q_0
q_1	q_2	q_0
$*q_2$	q_2	q_0

The symbol → in the table indicates that q_0 is the start state, and * indicates that q_2 is the final state.

23. Design a finite automata which will accept the language

L = {w ∈ (0,1)*/second symbol of w is '0' and fourth input is '1'}.

Ans: Here, we have to construct finite automata that will accept all the strings of which second symbol is 0 and fourth is 1. The finite automata for the given problem is shown below:

Hence, finite automata $M = \{Q, \Sigma, \delta, q_0, F\}$ will be

$Q = \{q_0, q_1, q_2, q_3, q_4, q_5\}$
$\Sigma = \{0, 1\}$
$q_0 = \{q_0\}$
$F = \{q_4\}$

The transition function δ is shown with the help of the following transition table:

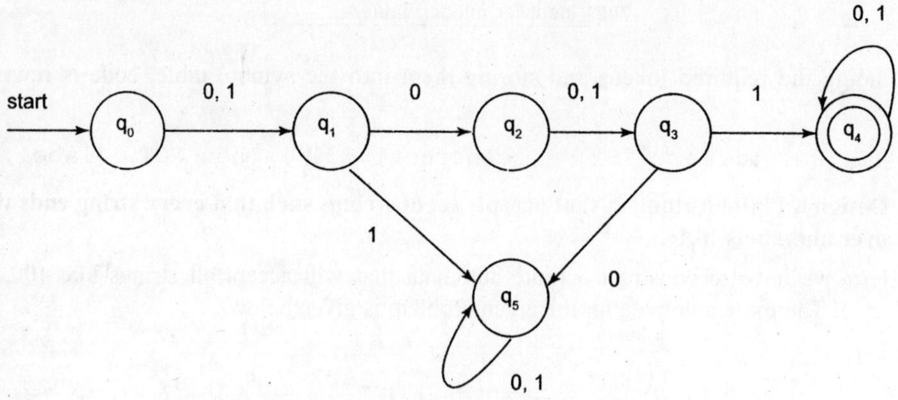

δ	0	1
→ q_0	q_1	q_1
q_1	q_2	q_5
q_2	q_3	q_3
q_3	q_5	q_4
*q_4	q_4	q_4
q_5	q_5	q_5

24. Construct a DFA for language over alphabet $\Sigma = \{a, b\}$ that will accept all strings beginning with 'ab'.

Ans: Here, we have to construct a DFA that will accept all strings beginning with ab like {ab, abb, abaab, ababb, abba, . . .}.

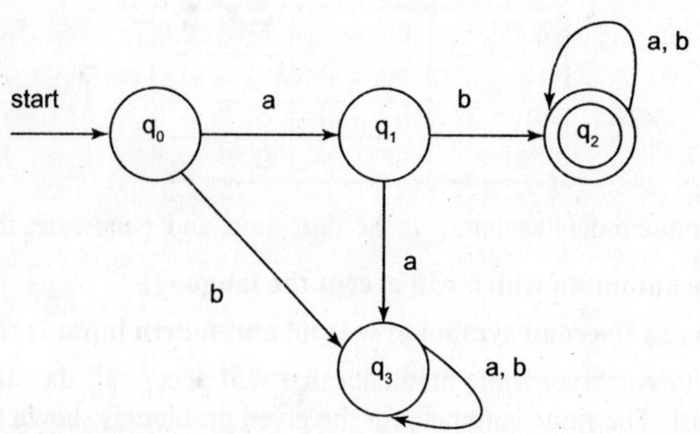

Hence, DFA $M = \{Q, \Sigma, \delta, q_0, F\}$ will be

$Q = \{q_0, q_1, q_2, q_3\}$
$\Sigma = \{a, b\}$
$q_0 = \{q_0\}$
$F = \{q_2\}$

The transition function δ is shown with the help of the following transition table:

δ	a	b
$\rightarrow q_0$	q_1	q_3
q_1	q_3	q_2
$*q_2$	q_2	q_2
q_3	q_3	q_3

25. Convert the following NFA into DFA.

$M = \{\{q_0, q_1\}, \{0,1\}, \delta, q_0 \{q_1\}\}$ and δ is

States	Inputs	
	0	1
$\rightarrow q_0$	$\{q_0, q_1\}$	$\{q_1\}$
q_1	ϕ	$\{q_0, q_1\}$

Ans: We will first draw the NFA according to the given transition table, as shown below:

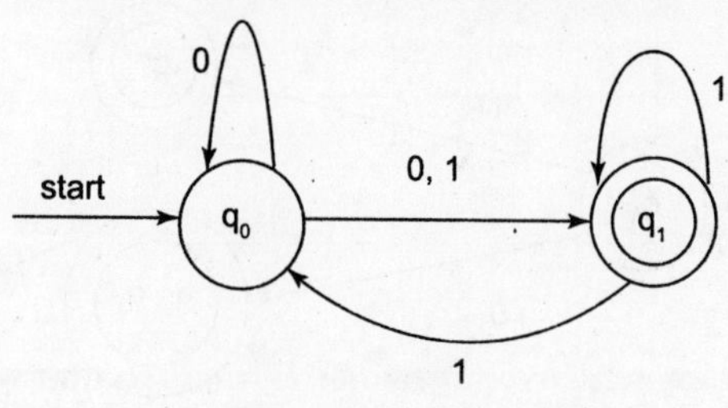

Now, we convert the NFA into DFA by following the given steps:

Step 1: Find all the transitions from initial state q_0 for every input symbol, that is, $\Sigma = \{0, 1\}$. If we get a set having more than one state for a particular input, then we consider that set as new single state. From the given transition table, it is clear that

$\delta(q_0, 0) \rightarrow \{q_0, q_1\}$, that is, q_0 transits to both q_0 and q_1 for input 0. (1)
$\delta(q_0, 1) \rightarrow \{q_1\}$, that is, for input 1, q_0 transits to q_1. (2)
$\delta(q_1, 0) \rightarrow \phi$, that is, for input 0, there is no transition from q_1. (3)
$\delta(q_1, 1) \rightarrow \{q_0, q_1\}$, that is, q_1 transits to both q_0 and q_1 for input 1. (4)

Step 2: In step 1, we have got a new state $\{q_0, q_1\}$. Now step 1 is repeated for this new state only, that is,

$\delta(\{q_0, q_1\}, 0) \rightarrow \delta(q_0, 0) \cup \delta(q_1, 0)$ (A)
Since $\delta(q_0, 0) \rightarrow \{q_0, q_1\}$ (from equation (1))
And $\delta(q_1, 0) \rightarrow \phi$ (from equation (3))

Therefore, equation (A) now becomes

$\delta(\{q_0, q_1\}, 0) \rightarrow \delta(q_0, 0) \cup \delta(q_1, 0) \rightarrow \{q_0, q_1\} \cup \phi \rightarrow \{q_0, q_1\}$ (5)

Now, consider $\delta(\{q_0, q_1\}, 1) \rightarrow \delta(q_0, 1) \cup \delta(q_1, 1)$ (B)

Since $\delta(q_0, 1) \rightarrow \{q_1\}$ (from equation (2))

And $\delta(q_1, 1) \rightarrow \{q_0, q_1\}$ (from equation (4))

Therefore, equation (B) now becomes

$\delta(\{q_0, q_1\}, 1) \rightarrow \delta(q_0, 1) \cup \delta(q_1, 1) \rightarrow \{q_0\} \cup \{q_0, q_1\} \rightarrow \{q_0, q_1\}$ (6)

Now, based on equations (1)–(6), we will construct the following transition table:

States	Inputs	
	0	**1**
$\rightarrow\{q_0\}$	$\{q_0, q_1\}$	$\{q_1\}$
$\{q_1\}$	ϕ	$\{q_0, q_1\}$
$\{q_0, q_1\}$	$\{q_0, q_1\}$	$\{q_0, q_1\}$

Since the starting state of given NFA is q_0, it will also be the starting state for DFA. Moreover, q_1 is the final state of NFA; therefore, we have to consider all those set of states containing q_1 as the member. All such sets will become the final states of DFA. Thus, F for the resultant DFA is:

$F = \{\{q_1\}, \{q_0, q_1\}\}$

The equivalent DFA for the given NFA is as follows:

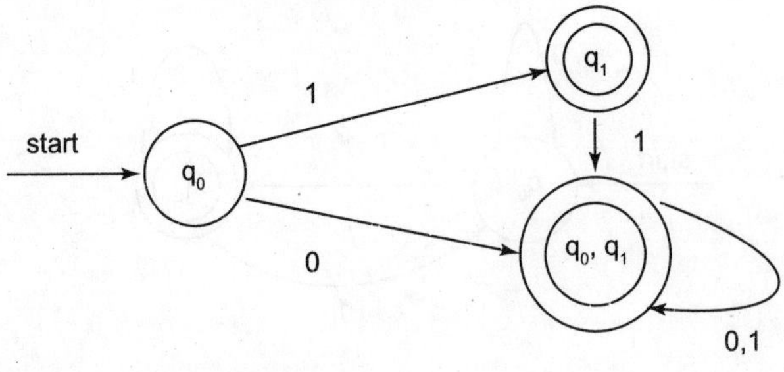

Now, we will relabel the DFA as follows:

$q_0 \rightarrow A$

$q_1 \rightarrow B$

$\{q_0, q_1\} \rightarrow C$

The transition table now becomes

States	Inputs	
	0	**1**
$\rightarrow$ A	C	B
*B	–	C
*C	C	C

The equivalent DFA after relabeling is as follows:

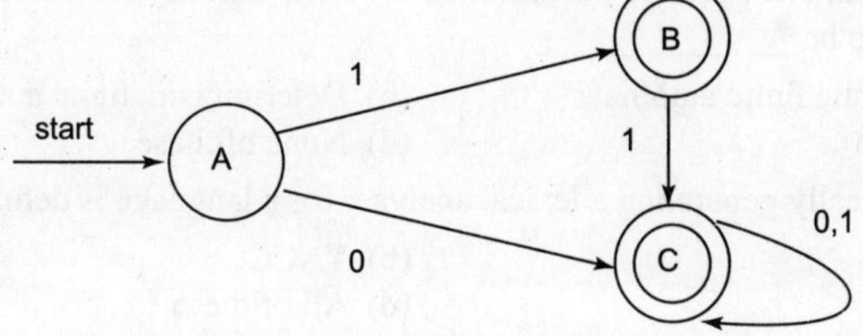

26. Convert the given regular expression into NFA.

(a/b) * a (a/b)

Ans:

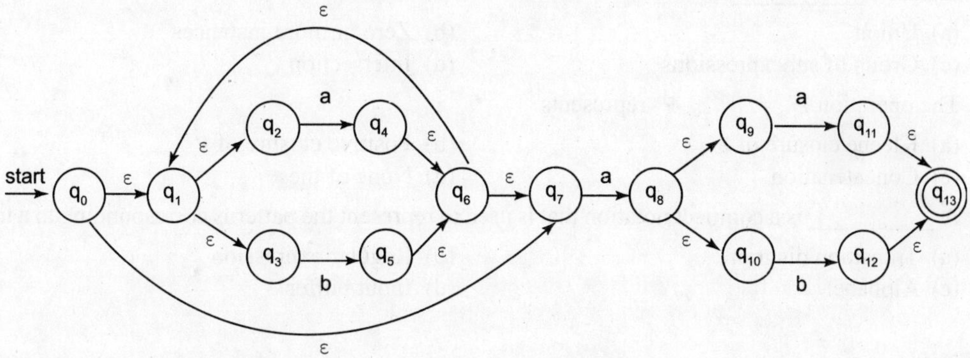

Multiple-Choice Questions

1. A __________ acts as an interface between the source program and the rest of the phases of compiler.

   (a) Semantic analyzer (b) Parser
   (c) Lexical analyzer (d) Syntax analyzer

2. Which of these tasks are performed by the lexical analyzer?

   (a) Stripping out comments and whitespace
   (b) Correlating error messages with the source program
   (c) Performing the expansion of macros
   (d) All of these

3. A __________ is any finite set of strings over some specific alphabet.

   (a) Sentence (b) Word
   (c) Language (d) Character class

4. If zero or more symbols are removed from the end of any string s, a new string is obtained known as a __________ of string s.

   (a) Prefix (b) Suffix
   (c) Substring (d) Subsequence

5. If we have more than one possible transition on the same input symbol from some state, then the recognizer is said to be __________.
   (a) Non-deterministic finite automata
   (b) Deterministic finite automata
   (c) Finite automata
   (d) None of these
6. A tool for automatically generating a lexical analyzer for a language is defined as __________.
   (a) Lex
   (b) YACC
   (c) Handler
   (d) All of these
7. `For A = 10 to 50 do`, in the given code, `A` is defined as a/an __________.
   (a) Constant
   (b) Identifier
   (c) Keyword
   (d) Operator
8. The language for C identifiers can be described as: `letter_(letter_|digit)`*, here * indicates __________.
   (a) Union
   (b) Zero or more instances
   (c) Group of subexpressions
   (d) Intersection
9. The operation $P^* = \cup_{i=0}^{\infty} P^i$ represents
   (a) Kleene closure of P
   (b) Positive closure of P
   (c) Concatenation
   (d) None of these
10. A __________ is a compact notation that is used to represent the patterns corresponding to a token.
    (a) Transition diagram
    (b) Regular expression
    (c) Alphabet
    (d) Input buffer

Answers

1. (c) 2. (d) 3. (c) 4. (a) 5. (a) 6. (a) 7. (b) 8. (b) 9. (a) 10. (b)

3

Specification of Programming Languages

1. Explain context-free grammar (CFG) and its four components with the help of an example.

Ans: The context-free grammar (CFG) was developed by Chomsky in 1965. A CFG is used to specify the syntactic structure of a programming language constructs like expressions and statements. The CFG is also known as **Backus-Naur Form (BNF)**. A CFG comprises four components, namely, *non-terminals*, *terminals*, *productions,* and *start symbol.*

- The **non-terminals** (also known as **syntactic variables**) represent the set of strings in a language.
- The **terminals** (also known as **tokens**) represent the symbols of the language.
- The **productions** or the rewriting rules represent the way in which the terminals and non-terminals can be joined to form a string. A production is represented in the form of A $\rightarrow$ α. This production includes a single non-terminal A, known as the left hand side or head of the production, an arrow, and a string of terminals and/or non-terminals α, known as the right hand side or body of the production. The components of the body represent the way in which the strings of the non-terminal at the head can be constructed. Productions of the start symbol are always listed first.
- A single non-terminal is chosen as the **start symbol** which represents the language that is generated from the grammar.

Formally, CFG can be represented as:

```
G = {V, T, P, S}
```

where

V is a finite set of non-terminals,
T is a finite set of terminals,
P is a finite set of productions,
S is the start symbol.

For example, consider an **if-else** conditional statement which can be represented as:

```
if (expression) statement else statement
```

The production for this statement is written as follows:

```
stmnt → if (expr) stmnt else stmnt
```

where `stmnt` is a variable used to denote `statement` and `expr` is a variable used to denote `expression`. Here, `expr` and `stmnt` are non-terminals, and the keywords **`if`** and **`else`** and the parenthesis are terminals. The arrow (→) can be read as 'can have the form'.

2. Consider the following grammar for arithmetic expressions and write the precise form of CFG using the shorthand notations.

```
statement  → statement + term
statement  → term
term       → term * factor
term       → factor
factor     → (statement)
factor     → id
```

Ans: The various shorthand notations used in grammars are as follows:

- The symbols used as non-terminals include uppercase starting alphabets (A, B, C, . . .). The lower-case names like expression, terms, factors, etc., are mostly represented as E, T, F, respectively, and letter S mostly used as the start symbol.
- The symbols used as terminals include lowercase starting alphabets (a, b, c, . . .), arithmetic operators (/, *, +, -), punctuation symbols (parenthesis, comma), and numbers (0, 1, . . . , 9). Lowercase alphabets like u, v, . . . , z are considered as strings of terminals. The boldface strings like **id** or **if** are also considered as terminals.
- Ending uppercase alphabets like X, Y, Z are used to represent either terminals or non-terminals.
- Lowercase Greek letters like α, β, γ are considered as string of terminal and non-terminals. A generic production can hence be represented as $A \rightarrow \alpha$, where A represents the left hand side of the production and α represents a string of grammar symbols (the right hand side of the production). A set of productions $A \rightarrow \alpha_1, A \rightarrow \alpha_2, \ldots, A \rightarrow \alpha_n$ can be represented as $A \rightarrow \alpha_1 | \alpha_2 | \ldots | \alpha_n$. The symbol '|' represent 'or'.

Considering these notations, the grammar can be written as follows:

```
S → S + T | T
T → T * F|F
F → (S) | id
```

3. What do you mean by derivation? What are its types? What are canonical derivations?

Ans: Derivation is defined as the replacement of non-terminal symbols in a particular string of terminals and non-terminals. The basic idea behind derivation is to apply productions repeatedly to expand the non-terminal symbol in that string. Consider the following productions:

```
E → (E) |-E |id
```

The single non-terminal E, at the head of the production, can be replaced by -E and it can be written as $E \Rightarrow -E$, which means "E derives -E". Similarly, E derives (E) can be written as $E \Rightarrow (E)$. The symbol $\Rightarrow$ means derives in one step. A sequence of replacements like $E \Rightarrow -E \Rightarrow -(E) \Rightarrow -(id)$ is called the derivation of -(id) from E. This denotes derivation in zero or more steps. The symbol $\overset{*}{\Rightarrow}$ is used to denote the derivation in zero or more steps. If there is a derivation $S \overset{*}{\Rightarrow} \alpha$ and S is the start symbol of a grammar G, then α is known as the **sentential form** of G. The symbol $\overset{+}{\Rightarrow}$ is used to denote derivation in one or more steps.

Based on the order of replacement of the non-terminals, derivation can be classified into two types, namely, *leftmost derivation* and *rightmost derivation*. In **leftmost derivation**, the leftmost non-terminal in each sentential is replaced with the equivalent production's right hand side. The leftmost derivation for $\alpha \Rightarrow \beta$ is represented as $\alpha \underset{lm}{\Rightarrow} \beta$.

In **rightmost derivation,** the rightmost non-terminal in each sentential is replaced with the equivalent production's right hand side. The rightmost derivation for $\alpha \Rightarrow \beta$ is represented as $\alpha \underset{rm}{\Rightarrow} \beta$.

For example, consider the following grammar:

```
S → XY
X → xxX
Y → Yy
X → ∈
Y → ∈
```

The leftmost derivation can be written as:

$$S \underset{lm}{\Rightarrow} XY \underset{lm}{\Rightarrow} xxXY \underset{lm}{\Rightarrow} xxY \underset{lm}{\Rightarrow} xxYy \underset{lm}{\Rightarrow} xxy$$

The rightmost derivation can be written as:

$$S \underset{rm}{\Rightarrow} XY \underset{rm}{\Rightarrow} XYy \underset{rm}{\Rightarrow} Xy \underset{rm}{\Rightarrow} xxXy \underset{rm}{\Rightarrow} xxy$$

The rightmost derivations are also known as **canonical derivations.**

4. Write a grammar to generate a palindrome.

Ans: A string that is read same in either direction is known as palindrome. For example, the string *madam* is a palindrome. Consider the following productions using which a palindrome, `ababa`, can be generated.

```
S → aSa
S → bSb
S → a
S → b
S → ∈
```

Hence, the string `ababa` can be generated as follows:

$$S \Rightarrow aSa \Rightarrow abSba \Rightarrow ababa$$

5. Define the term sententials. What is a context-free language? When two languages are said to be equivalent?

Ans: The intermediate strings in a derivation that consists of terminals and non-terminals are called **sententials**. The sentential form that occur in a leftmost derivation is known as **left sentential form** and that occur in a rightmost derivation is known as **right sentential form**.

A sentential form that contains only terminals is called a **sentence** of a grammar G. A set of sentences generated by a grammar forms the language, which is known as **context-free language**. The grammars are said to be equivalent if two grammars generate the same language.

6. What is an ambiguous grammar? Specify the demerits of ambiguous grammar. Explain with the help of an example how ambiguity can be removed.

Ans: An **ambiguous grammar** is a grammar that generates more than one leftmost or rightmost derivation for some sentences. For example, consider the following grammar to produce the string `id - id/id`.

```
E → E - E | E/E
E → id
```

This grammar is ambiguous since it generates more than one leftmost derivation.
One derivation is as follows:

```
E → E - E
E → id - E/E
E → id - id/E
E → id - id/id
```

Another derivation is as follows:

```
E → E/E
E → E - E/E
E → id - E/E
E → id - id/E
E → id - id/id
```

The demerit of an ambiguous grammar is that it generates more than one parse tree for a sentence and, hence, it is difficult to choose the parse tree to be evaluated.

Ambiguity in grammars can be removed by rewriting the grammar. While rewriting the grammar, two concepts must be considered, namely, *operator precedence* and *associativity*.

- **Operator precedence:** Operator precedence indicates the priority given to the arithmetic operators like /, *, +, –. The operators, * and /, have higher precedence than + and –. Hence, a string `id - id/id` is interpreted as `id - (id/id)`.
- **Associativity of operators:** The associativity of operators involves choosing the order in which the arithmetic operators having the same precedence occur in a string. The arithmetic operators follow left to right associativity. Hence, a string `id + id - id` is interpreted as `(id + id) - id`. Some other operators like exponentiation and assignment operator = follow right to left associativity. Hence, a string `id↑id↑id` is interpreted as `id↑(id↑id)`.

7. Discuss dangling else ambiguity.

Ans: Dangling else ambiguity is a form of ambiguity that occurs in grammar while representing conditional constructs of programming language. For example, consider the following grammar for the conditional statements:

```
statement → if condition then statement
statement → if condition then statement else statement
statement → other statement
```

Now, consider the following string:

```
if C1 then if C2 then S1 else S2
```

Since this string generates two parse trees as shown in Figure 3.1, the grammar is said to be ambiguous.

This ambiguity can be eliminated by matching each **else** with its just preceding unmatched **then**. It generates a parse tree for the string that relates each **else** with its closest previous unmatched **then**. The unambiguous grammar is written as follows:

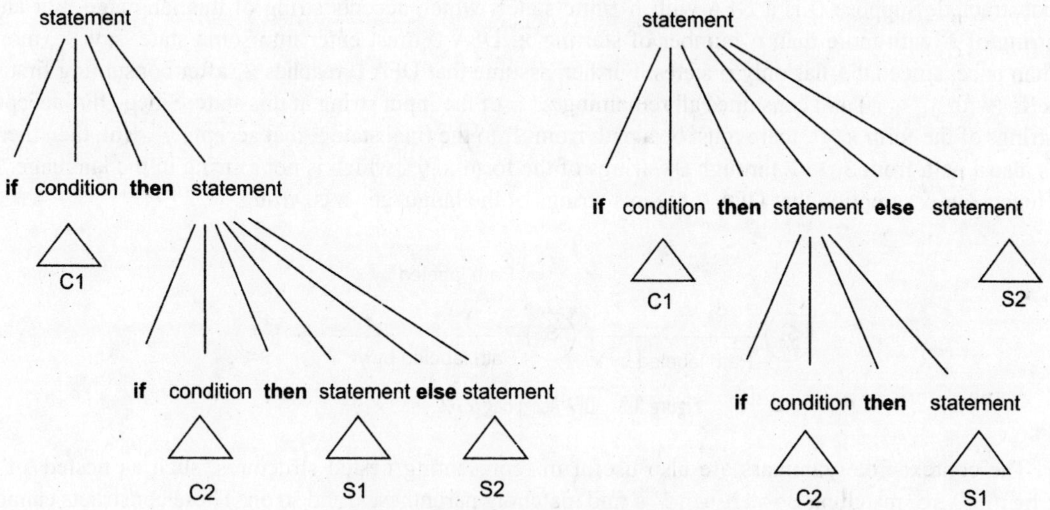

Figure 3.1 Parse Trees for Ambiguous Grammar

```
          statement → matched statement | unmatched statement
  matched statement → if condition then matched statement else
                      matched statement | other statement
unmatched statement → if condition then statement
                      |if condition then matched statement else
                      unmatched statement
```

8. What are the advantages of context-free grammar?

Ans: The advantages of a context-free grammar are as follows:

- It gives a simple and easy to understand syntactic specification of the programming language.
- It can construct an efficient parser.
- Imparting structure to a program, a grammar helps to translate it to an object code and also helps in the detection of errors in the program.

9. What are the capabilities of CFG?

Ans: Any syntactic construct that can be represented using a regular expression can also be represented using a context-free grammar, but not vice-versa. Hence, a context-free grammar is more capable of representing a language than regular expressions. Consider the regular expression `(x|y) * xyy`. The context-free grammar given below generates the same language, with the string ending with `xyy`.

```
S → xS |yS
S → xA
A → yB
B → yC
C → ∈
```

Now, consider a language $L = \{x^m y^m | m >= 1\}$ described by context-free grammar. Assume that this language can be described by regular expression. It means a DFA for this language can be constructed. Suppose D is a DFA with n finite states, which accepts string of this language. For any string of L with more than n number of starting x, DFA D must enter into some state, say S_i, more than once, since DFA has only n states. Further, assume that DFA D reaches S_i after consuming first j x's (with j < m) and consumes all remaining x's of the input string at this state. Since DFA accepts strings of the form $x^m y^m$, there must be a path from S_i to the final state F that accepts y^m. But, then there is also a path from S_0 to F through S_i strings of the form $x^j y^m$, which is not a string in the language L. Hence, our assumption that DFA D accepts strings of the language L is wrong.

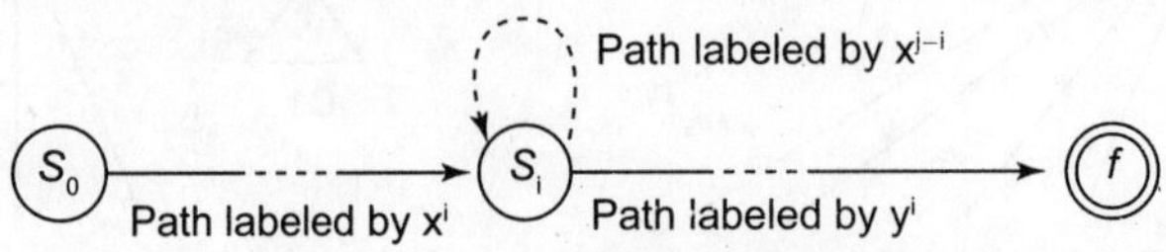

Figure 3.2 DFA Accepting $x^i y^j$ and $x^j y^i$

The context-free grammars are also useful in representing nested structures, such as nested if-then-else, matching begin-end's and matching parentheses, and so on. These constructs cannot be represented using regular expressions.

10. Why the use of CFG is not preferred over regular expressions for defining the lexical syntax of a language?

Ans: Regular expressions are preferred over CFG to describe the lexical syntax of a language due to the following reasons:

- Regular expressions provide a simple notation for tokens as compared to grammars.
- The lexical rules provided by regular expressions are quite simple, and hence, a powerful notation like CFG is not required.
- Regular expressions are used to construct more efficient lexical analyzers.
- The syntactic structure of a language when divided into lexical and non-lexical parts provides an easy way to modularize the front end of a compiler.
- The lexical constructs like identifiers, constants, keywords, etc., can be easily described using regular expressions.

11. What do you mean by a left recursive grammar? Write an algorithm to eliminate left recursion.

Ans: For a grammar G, if there exists a derivation $A \overset{+}{\Rightarrow} A\alpha$ for some string α, then the grammar is said to be **left recursive**. Left recursion causes problem while designing parsers (parsers are discussed in the next chapter). When the parse tree for a left recursive grammar is constructed, the process gets into an infinite loop. This looping results in an invalid string.

Left recursion can be eliminated by rewriting the offending production. Consider the production, $E \rightarrow E + T \mid T$, where the non-terminal on the left hand side of the production is the same as the leftmost symbol on the right hand side. Now, if in the production we try to expand E, it will eventually result in again expanding E without taking any input. So, left recursion can be eliminated by replacing $E \rightarrow E + T \mid T$ with $E \rightarrow TE'$ and $E' \rightarrow + TE \mid \in$. This process eliminates the immediate left recursion; however, eliminating left recursion from the grammar involving derivations of two or more steps is not possible. Hence, an algorithm is designed for such derivations as shown in Figure 3.3. This algorithm is suitable for grammars with no cycles or $\in$ productions.

```
Step 1: Arrange the non-terminals of a grammar G in the order, X1, X2,
        . . . , Xn.
Step 2: for i = 1 to m
        begin
           for j = 1 to i - 1
              begin
                 replace each production of Xi → Xj γ by the
                 productions Xi → α1 γ | α2 γ |. . .| αm γ |, where
                 Xj → α1 | α2 |. . .| αm are all current Xj productions
              end
              eliminate immediate left recursion from the Xi
              productions
        end
```

Figure 3.3 Algorithm for Eliminating Left Recursion

12. Explain left factoring with the help of an example.

Ans: When productions contain common prefixes, it is difficult to choose a valid rule while applying the production. Hence, in such a situation left factoring is used. For example, consider a grammar with two productions:

$E \rightarrow \alpha\ T_1\ |\alpha\ T_2$

The common prefix α in this grammar makes it difficult to choose between $\alpha\ T_1$ and $\alpha\ T_2$ for expanding E. Hence, the grammar is left factored and the productions can be rewritten as follows:

$E \rightarrow \alpha\ E'$
$E' \rightarrow T_1\ |T_2$

Now, after expanding E to $\alpha\ E'$, we can expand $E' \rightarrow T_1$ or T_2 by seeing the input derived from α. An algorithm for left factoring a grammar is shown in Figure 3.4.

```
begin
   for a grammar G with a non-terminal X, find the longest prefix α
   common to two or more of its alternatives.
      If α ≠ ∈, then replace all of the X-productions
         X → αβ1 | αβ2 |. . .| αβn |γ, with
            X → αX' | γ
            X' → β1 |β2 |. . .|βn
   where
      γ specifies the alternatives that do not begin with α and X' is
      a new non-terminal.
         Repeat this process until no two alternatives for a non-
         terminal have a common prefix.
end
```

Figure 3.4 Algorithm for Left Factoring

13. Consider the following grammar:

S → A | ∧ | (T)
T → T, S |S

In the above grammar, find the leftmost and rightmost derivations for

(a) (A, (A,A))
(b) (((A,A), ∧, (A)),A).

Ans: (a) The leftmost derivation for the string (A, (A,A)) can be written as follows:

$$S \underset{lm}{\Rightarrow} (T) \underset{lm}{\Rightarrow} (T,S) \underset{lm}{\Rightarrow} (S,S) \underset{lm}{\Rightarrow} (A,S)$$
$$\underset{lm}{\Rightarrow} (A, (T)) \underset{lm}{\Rightarrow} (A, (T,S)) \underset{lm}{\Rightarrow} (A, (S,S))$$
$$\underset{lm}{\Rightarrow} (A, (A,S)) \underset{lm}{\Rightarrow} (A, (A,A))$$

The rightmost derivation for the string (A, (A,A)) can be written as follows:

$$S \underset{rm}{\Rightarrow} (T) \underset{rm}{\Rightarrow} (T, S) \underset{rm}{\Rightarrow} (T, (T)) \underset{rm}{\Rightarrow} (T, (T,S))$$
$$\underset{rm}{\Rightarrow} (T, (T,A)) \underset{rm}{\Rightarrow} (T, (S,A)) \underset{rm}{\Rightarrow} (T, (A,A))$$
$$\underset{rm}{\Rightarrow} (S, (A,A)) \underset{rm}{\Rightarrow} (A, (A,A))$$

(b) The leftmost derivation for the string (((A,A), ∧, (A)),A) can be written as follows:

$$S \underset{lm}{\Rightarrow} (T) \underset{lm}{\Rightarrow} (T,S) \underset{lm}{\Rightarrow} (S,S) \underset{lm}{\Rightarrow} ((T),S) \underset{lm}{\Rightarrow} ((T,S),S)$$
$$\underset{lm}{\Rightarrow} ((T,S,S),S) \underset{lm}{\Rightarrow} ((S,S,S),S) \underset{lm}{\Rightarrow} (((T),S,S),S)$$
$$\underset{lm}{\Rightarrow} (((T,S),S,S),S) \underset{lm}{\Rightarrow} (((S,S),S,S),S) \underset{lm}{\Rightarrow} (((A,S),S,S),S)$$
$$\underset{lm}{\Rightarrow} (((A,A),S,S),S) \underset{lm}{\Rightarrow} (((A,A),\wedge,S),S)$$
$$\underset{lm}{\Rightarrow} (((A,A),\wedge,(T)),S) \underset{lm}{\Rightarrow} (((A,A),\wedge,(S)),S)$$
$$\underset{lm}{\Rightarrow} (((A,A),\wedge,(A)),S) \underset{lm}{\Rightarrow} (((A,A),\wedge,(A)),A)$$

The rightmost derivation for the string (((A,A), ∧, (A)),A) can be written as follows:

$$S \underset{rm}{\Rightarrow} (T) \underset{rm}{\Rightarrow} (T,S) \underset{rm}{\Rightarrow} (T,A) \underset{rm}{\Rightarrow} (S,A) \underset{rm}{\Rightarrow} ((T),A)$$
$$\underset{rm}{\Rightarrow} ((T,S),A) \underset{rm}{\Rightarrow} ((T,(T)),A) \underset{rm}{\Rightarrow} ((T,(S)),A) \underset{rm}{\Rightarrow} ((T,(A)),A)$$
$$\underset{rm}{\Rightarrow} ((T,S,(A)),A) \underset{rm}{\Rightarrow} ((T,\wedge,(A)),A) \underset{rm}{\Rightarrow} ((S,\wedge,(A)),A)$$
$$\underset{rm}{\Rightarrow} (((T),\wedge,(A)),A) \underset{rm}{\Rightarrow} (((T,S),\wedge,(A)),A)$$
$$\underset{rm}{\Rightarrow} (((T,A),\wedge,(A)),A) \underset{rm}{\Rightarrow} (((S,A),\wedge,(A)),A) \underset{rm}{\Rightarrow} (((A,A),\wedge,(A)),A)$$

14. Prove the grammar is ambiguous.

```
E → E + E | E * E | (E) |id
```

Ans: Ambiguity of a grammar can be proved by showing that it generates more than one leftmost derivation or rightmost derivation for some string or sentence. Now, consider a string `(id) + id * id`. One leftmost derivation for this string is as follows:

```
E → E + E
E → (E) + E
E → (id) + E
E → (id) + E * E
E → (id) + id * E
E → (id) + id * id
```

Another derivation for the same string is as follows:

```
E → E * E
E → E + E * E
E → (E) + E * E
E → (id) + E * E
E → (id) + id * E
E → (id) + id * id
```

Since there are more than one leftmost derivation for the sample string, the given grammar is ambiguous.

15. Eliminate left recursion for the following grammar:

```
S → S + E | E
E → E * F | F
F → (S) | id
```

Ans: The left recursion can be eliminated by using the following productions:

```
S → E S'
S'→ + E S'|∈
E → F E'
E'→ *F E'|∈
F → (S) |id
```

16. Perform left factoring for the following grammar:

```
A → aBcC | aBb | aB | a
B → ∈
C → ∈
```

Ans: Applying left factoring, the grammar can be written as:

```
A → aA'
A'→ BcC |Bb |B
B → ∈
C → ∈
```

Multiple-Choice Questions

1. Which of the following grammar is also known as Backus-Naur form?
   (a) Regular
   (b) Context-free
   (c) Context-sensitive
   (d) None of these
2. In `G = {V, T, P, S}` representation of context-free grammar, 'V' stands for _________.
   (a) A finite set of terminals
   (b) A finite set of non-terminals
   (c) A finite set of productions
   (d) Is the start symbol
3. Which of these statements are correct for the productions in context-free grammar?
   (a) Productions represent the way in which the terminals and non-terminals can be joined to form a string.
   (b) The left hand side of the production contains a single non-terminal.
   (c) The right hand side of the production contains a string of terminals and/or non-terminals.
   (d) All of these
4. _________ is defined as the replacement of non-terminal symbols in a particular string of terminals and non-terminals.
   (a) Production
   (b) Derivation
   (c) Sentential form
   (d) Left factoring
5. In a derivation _________ are the intermediate strings that consists of terminals and non-terminals.
   (a) Sententials
   (b) Context-free language
   (c) Context-sensitive language
   (d) None of these
6. A grammar generating more than one derivation for some sentences is known as _________.
   (a) Regular
   (b) Context-free
   (c) Context-sensitive
   (d) Ambiguous
7. A grammar contains _________.
   (a) A non-terminal `V` that can be present in any sentential form
   (b) A non-terminal `V` that cannot derive any string of terminals
   (c) ε as the only symbol in the left hand side of production
   (d) None of these
8. Which of these are also known as canonical derivations?
   (a) Leftmost derivations
   (b) Rightmost derivations
   (c) Sentential form
   (d) None of these
9. Which of these statements is correct?
   (a) Sentence of a grammar is a sentential form without any terminals.
   (b) Sentence of a grammar should be derivable from the `start` state.
   (c) Sentence of a grammar is a sentential form with no non-terminals.
   (d) All of these

10. Consider a grammar: $A \rightarrow \alpha S_1 \mid \alpha S_2$, the left factored productions for this grammar are:

(a) $A' \rightarrow \alpha A$
$A \rightarrow S_1 \mid S_2$

(b) $A \rightarrow \alpha A'$
$A' \rightarrow \alpha S_1 \mid \alpha S_2$

(c) $A \rightarrow \alpha A'$
$A' \rightarrow S_1 \mid S_2$

(d) None of these

Answers

1. (b) 2. (b) 3. (d) 4. (b) 5. (a) 6. (d) 7. (a) 8. (b) 9. (c) 10. (c)

4

Basic Parsing Techniques

1. Define parsing. What is the role of a parser?

Ans: Parsing (also known as **syntax analysis**) can be defined as a process of analyzing a text which contains a sequence of tokens, to determine its grammatical structure with respect to a given grammar.

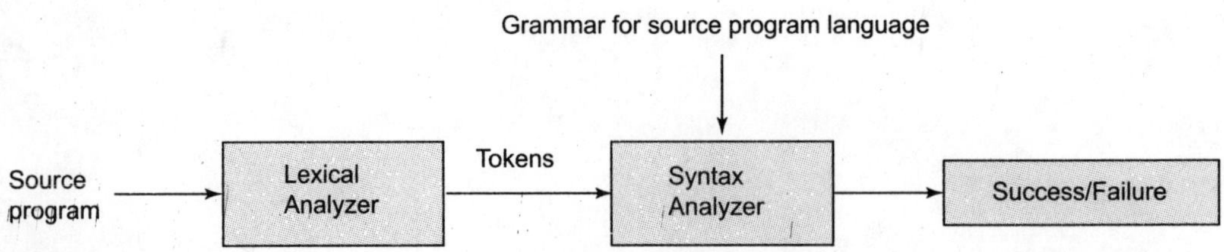

Figure 4.1 Parsing

Depending upon how the parse tree is built, parsing techniques are classified into three general categories, namely, *universal parsing*, *top-down parsing,* and *bottom-up parsing*. The most commonly used parsing techniques are top-down parsing and bottom-up parsing. Universal parsing is not used as it is not an efficient technique. The hierarchical classification of parsing techniques is shown in Figure 4.2.

Role of a parser: A parser receives a string of tokens from lexical analyzer and constructs a parse tree if the string of tokens can be generated by the grammar of the source language; otherwise, it reports the syntax errors present in the source string. The generated parse tree is passed to the next phase of the compiler, as shown in Figure 4.3.

The role of parser is summarized as follows:

- Perform context-free syntax analysis.
- Guides context-sensitive analysis.
- Generate an intermediate code.
- Report syntax errors in an intelligible manner.
- Attempts error correction.

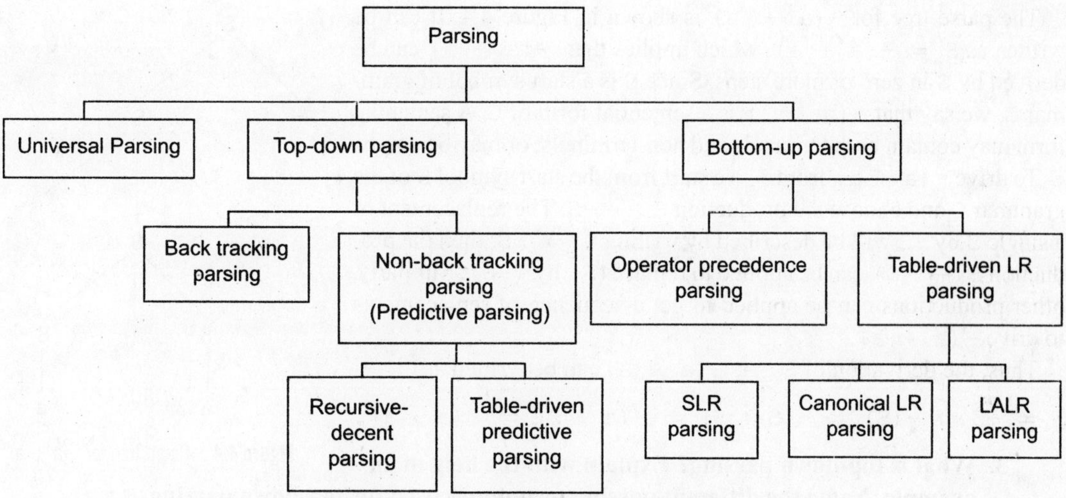

Figure 4.2 Classification of Parsing Techniques

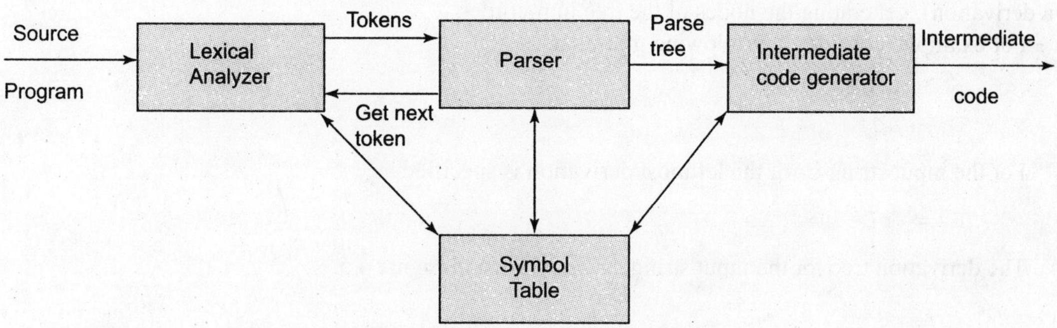

Figure 4.3 Position of a Parser in Compiler Model

2 Define parse tree. Construct the derivation of a parse tree.

Ans: A **parse tree** or **concrete syntax tree** is a tree data structure that is ordered and rooted and represents the syntactic structure (or grammatical structure) of a string. It can also be defined as a graphical representation of a derivation that shows the order in which grammar rules (productions) are applied to replace the non-terminals. A non-terminal S present in the head of the production is used to label the interior node and all of its children are labeled from left to right by the production symbols so that S can be replaced during the derivation.

Derivation of a parse tree: Derivations help in constructing a precise parse tree where productions are considered as rewriting rules. At each rewriting step in a derivation, a non-terminal is selected to be replaced and a production is taken for the derivational view with that non-terminal as head.

For example, consider the following grammar G:

S → S + S|S * S|(S)| - S|a

The parse tree for `-(a + a)` is shown in Figure 4.4. It can be written as `S` $\overset{*}{\Rightarrow}$ `-(a + a)`, which implies that `-(a + a)` can be derived by `S` in zero or more steps. Since `S` is a start symbol of grammar `G`, we say that `-(a + a)` is a sentential form of `G`. A sentential form may contain both terminals and non-terminals, or may be empty.

To drive `-(a + a)` from `S`, we start from the start symbol `S` of the grammar `G` and choose the production `S → -S`. The replacement of a single `S` by `-S` will be described by writing `S ⇒ -S`; then the production `S → (S)` can be applied to replace `(S)` by `(-S)`. Similarly, other productions can be applied to get a sequence of replacements to drive `-(a + a)`.

Thus, the derivation of `S → -(a + a)` can be written as

`S ⇒ -S ⇒ -(S) ⇒ -(S + S) ⇒ -(a + S) ⇒ -(a + a).`

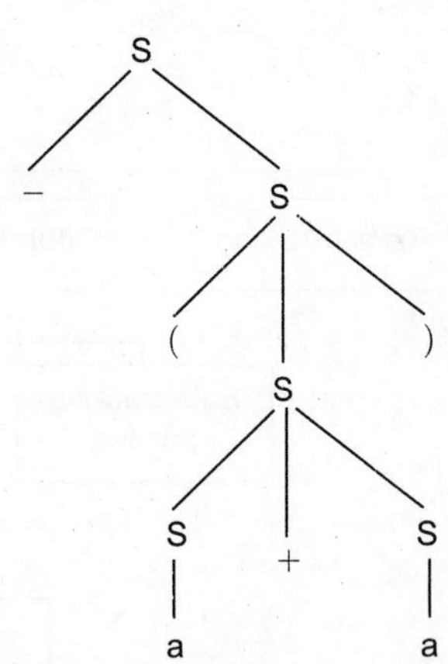

Figure 4.4 Parse Tree for –(a + a)

3. What is top-down parsing? Explain with the help of an example. Name the different parsing techniques used for top-down parsing.

Ans: Top-down parsing is a strategy to find the leftmost derivation of an input string. In top-down parsing, the parse tree is constructed starting from the root and proceeding toward the leaves (similar to a derivation), generating the nodes of the tree in preorder.

For example, consider the following grammar:

```
E → cDe
D → ab|a
```

For the input string `cae`, the leftmost derivation is specified as:

```
E ⇒ cDe ⇒ cae
```

The derivation tree for the input string `cae` is shown in Figure 4.5.

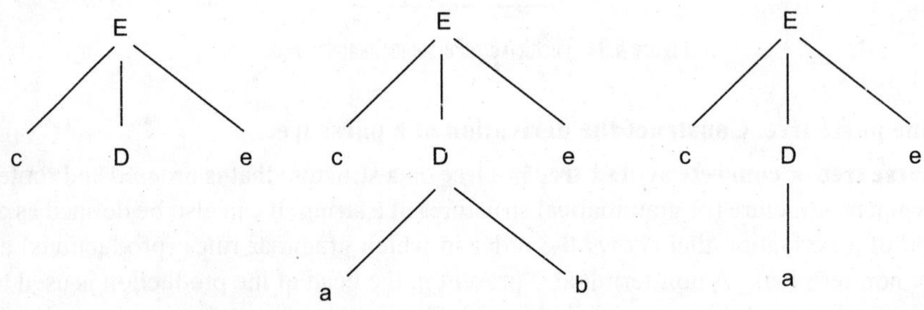

Figure 4.5 Parse Tree for `E → cae`

Top-down parsing techniques:

- **Backtracking parsing:** In this technique, the parser initiates with the start symbol and applies the grammar productions until the target string is generated. The productions are applied according to the order in which they are specified in the grammar. If a choice leads to a dead end, then the parser backtracks to the last decision point, undo that decision, and try another production until the parser finds a suitable production for matching the whole input or it runs out of choices.

- **Non-backtracking parsing (Predictive parsing):** Predictive parsing does not require backtracking in order to derive the input string. Predictive parsing is possible only for the class of `LL(k)` grammar (context-free grammar). The grammar should be free from left recursion and left factoring. Each non-terminal is combined with the next input signal to guide the parser to select the correct production rule that will lead the parser to match the complete input string.

There are two techniques of implementing top-down predictive parsers, namely, *recursive-decent* and *table-driven predictive parsings*.

4. Define recursive predictive parsing or predictive parsing.

Ans: It is a top-down parsing method, which consists of a set of mutually recursive procedures to process the input and handles a stack of activation records explicitly. The algorithm for predictive parsing is given in Figure 4.6.

```
Repeat
    Set A to the top of the stack and a the next input symbol
    If (A is a terminal or $) then
    If (A = a) then
        pop A from the stack and remove a from the input
    else
        /* error occurred */
    else /* A is a non-terminal */
        if (M[A,a] = A → B1B2B3...Bk) then
            /*M is the parsing table for grammar G*/
        Begin
            Pop A from the stack
            Push Bk, Bk-1,...,B1 onto the stack, B1 as top
        End
Until(A = $) /* stack becomes empty */
```

Figure 4.6 Algorithm of Predictive Parsing

In predictive parsing, a parsing table is constructed. To construct the table, we need two functions, namely, `FIRST()` and `FOLLOW()`, that are associated with the grammar `G`. These two functions are used to fill the proper entries in the table for `G`, if such a parsing table for `G` exists. The algorithm to construct the predictive parsing table is given in Figure 4.7.

```
For each production X → α of grammar G do
Begin
    For each terminal a in FIRST(α)
        add X → α to M [X, a]
End
If (FIRST(α) contains ∈) then
    For each terminal b in FOLLOW (X) add X → α to M [X, b]
If (FIRST(α) contains ∈) And (FOLLOW(X) contains $) then
```

```
    add X → α to M [X, $]
If (M[X, a] contains no more productions) then
    Set M [X, a] to error
```

Figure 4.7 Algorithm to Construct Predictive Parsing Table

5. Explain the FIRST-FOLLOW procedure.

Ans: For a grammar G, we need two functions **FIRST** and **FOLLOW** to construct the predictive parsing table for G. The first and follow procedure permit us to select which production to apply, based on the next input symbol.

The FIRST(A) is the set of all the terminals that A can begin with. If $A \stackrel{*}{\Rightarrow} \in$, then ∈ is also in FIRST(A). To compute FIRST(A) for all grammar symbols, the following rules can be applied:

1. If A is a terminal, then FIRST(A) = {A}.
2. If A is a non-terminal and $A \rightarrow X_1X_2 \ldots X_k \ldots X_n$ is a production.
   - If FIRST(X_1) does not contain ∈, then FIRST(A) = FIRST(X_1).
   - If FIRST(X_1) contains ∈ and FIRST(X_2) does not contain ∈, then
     FIRST(A) = FIRST(X_2).
   - If FIRST(X_1) and FIRST(X_2) both contain ∈ and FIRST(X_3) does not contain ∈, then
     FIRST(A) = FIRST(X_3).

   In general,
     FIRST(A) = FIRST (X_k) if FIRST (X_1), FIRST (X_2) . . . FIRST (X_{k-1}) all contain ∈ and FIRST (X_k) does not contain ∈.
3. If there exists a production A → ∈, then add ∈ to FIRST (A).

FOLLOW(A) for a non-terminal A is the set of all the terminals that can follow A. The FOLLOW set does not contain ∈. To compute FOLLOW(A) for all grammar symbols A, these rules can be applied:

1. Set FOLLOW(S) = $, where S is the start symbol and the symbol $ indicates the end of input.
2. If there exists a production A → βXγ, where γ is not ∈, then everything in FIRST(γ), except ∈, is in FOLLOW(X).
3. If there exists a production A → βX, or A → βXγ, where FIRST(γ) contains ∈, then all in FOLLOW(A) is in FOLLOW(X).

6. Define LL(1) grammars with an example.

Ans: The first 'L' in LL (1) stands for left to right scanning of input, the second 'L' to construct a leftmost derivation, and '1' indicates at each step, parser considers only one symbol of lookahead to make parsing decision. LL (1) grammars are very well-liked because the corresponding LL parsers only require to look at the next token to make their parsing decision. The LL (1) grammar is always non-left-recursive and unambiguous.

For a grammar G if X → α | β are two dissimilar productions, then G is LL (1) if and only if the given conditions are satisfied:

- For no terminal b do both α and β derive strings that begin with b.
- Empty string can be derived by at most one of the production α and β.

- If $\beta \overset{*}{\Rightarrow} \in$, then α does not derive any string that starts with a terminal in FOLLOW(X). Similarly, if $\alpha \overset{*}{\Rightarrow} \in$, then β does not derive any string that starts with a terminal in FOLLOW(X).

For example, consider the following grammar to construct a LL(1) parse table:

```
S → aABb
A → c|∈
B → d|∈
```

First, find out the FIRST and FOLLOW sets of all nonterminals.

```
  FIRST(S)  = {a}
  FIRST(A)  = {c,∈}
  FIRST(B)  = {d,∈}
 FOLLOW(S)  = {$}, since S is the start symbol.
 FOLLOW(A)  = FIRST(Bb)
            = FIRST(B)  - {∈} U FIRST (b)
            = {d,∈} - {∈} U {b}
            = {d, b}
 FOLLOW(B)  = {b}
```

Now,

1. Considering the production S → aABb
   FIRST (S) = {a}
   Since it does not contain any ∈.
   So, parse table [S, a] = S → aABb (1)
2. Considering the production A → c
   FIRST(A) = FIRST(c) = {c}.
   Since it does not contain any ∈.
   So, parse table [A, c] = A → c (2)
3. Considering the production A → ∈
   FIRST(A) = FIRST(∈) = {∈}.
   Since it contains ∈. Thus, we have to find out FOLLOW(A).
   FOLLOW(A) = {d, b}
   So, parse table [A, d] = A → ∈
   Also, parse table [A, b] = A → ∈ (3)
4. Considering the production B → d
   FIRST(B) = FIRST(d) = {d}.
   Since it does not contain any ∈.
   So, parse table [B, d] = B → d (4)
5. Considering the production B → ∈
   FIRST(B) = FIRST(∈) = {∈}.
   Since it contain ∈. Thus, we have to find out FOLLOW(B).
   FOLLOW(B) = {b}
   So, parse table [B, b] = B → ∈ (5)

Thus, the resultant parse table, from (1), (2), (3), (4), and (5) is shown in Table 4.1

Table 4.1 LL(1) Parsing Table

	a	b	c	d	$
S	S → aABb				
A		A → ∈	A → c	A → ∈	
B		B → ∈		B → d	

7. Write down the algorithm for recursive-decent parsing. Explain with an example.

Ans: A recursive-decent parser is a collection of procedures one for each non-terminal. Starting from the start symbol, the parser continues its scanning until it stops and announces success if it scans entire input string. The algorithm for recursive-decent parsing is shown in Figure 4.8.

```
void X()
Begin
    Select an X-production, X → A1 A2 . . . An
    For(i = 1 to n) do
    Begin
        If(Ai is a non-terminal)
            call procedure Ai( )
        else if(Ai = a)/* a is the current input symbol*/
            lead the input to the next symbol
        else /* presence of some error */
    End
End
```

Figure 4.8 Algorithm of Recursive-Decent Parsing

For example, consider the following grammar:

X → aMb
M → cd|c

Input string is acb.

For constructing parse tree, starting from X, we check the available productions for X. As X has only one production, so we use it to spread X and obtain the tree shown in Figure 4.9(a). As the leftmost symbol matches with the first symbol of input string, we move toward the second symbol, which is M, so we expand M with first substitute M → cd to obtain the tree shown in Figure 4.9(b).

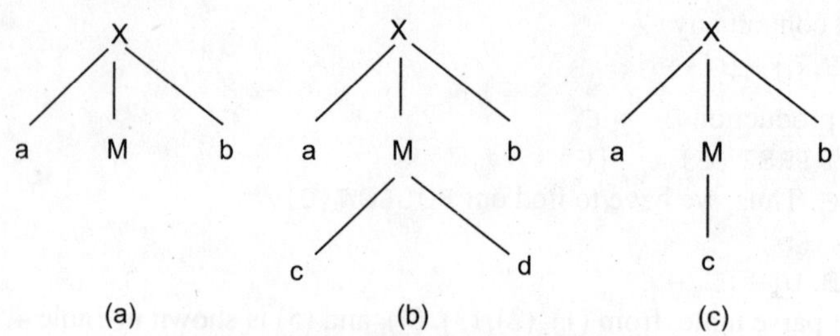

Figure 4.9 Parse Tree for Input String acb

Now, we get a match for the second symbol c, but as we proceed toward the third symbol a failure occurs because symbol b does not match with d. Now we go back to M and select second production of M → c, and prepare its parse tree shown in Figure 4.9(c). The leaf c matches with the second symbol of input string and leaf b with the third symbol of input string. Hence, we declare a successful parsing for the input string.

8. Explain the non-recursive predictive parsing or table-driven predictive parsing.

Ans: The non-recursive predictive parser explicitly maintains a stack instead of implicit stack that is maintained via recursive calls. The parser imitates a leftmost derivation. For an input a, the stack holds a sequence of grammar symbols β such that $S \underset{lm}{\Rightarrow} a\beta$.

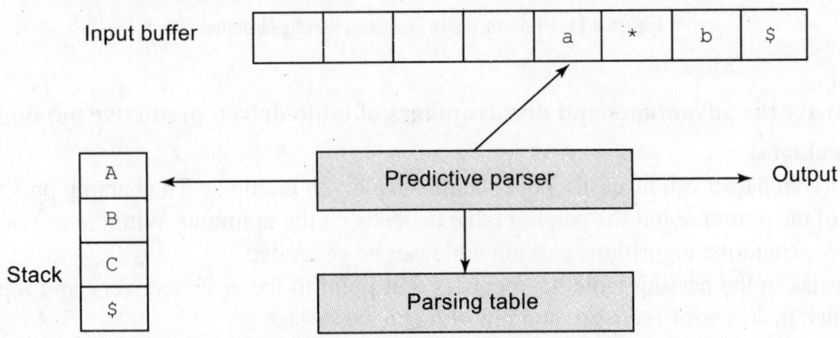

Figure 4.10 Non-recursive Predictive Parser

The non-recursive predictive parser contains:

- A stack containing a sequence of grammar symbol with $ at the bottom of the stack,
- An input buffer containing string to be parsed with an end marker $,
- A parsing table,
- An output stream.

The non-recursive predictive parser is controlled by a program that takes the top of the stack A, and the current input symbol a in consideration. If A is a non-terminal, then the parser chooses an A-production from parsing table M if there is an entry for M[A, a]. If A is a terminal, then it tries for a match between present input symbol a and the terminal A.

For a grammar G with a string d and a parsing table M, a leftmost derivation of d should exist if d is in L(G). Let us consider d$ in the input buffer and grammar start symbol S above $ is on the top of stack. So the predictive parser uses its parsing table to parse the input according to the following procedure shown in Figure 4.11.

```
Set instruction pointer to point the first symbol of d
Set top stack symbol to A
While(A ≠ $) do
Begin
    If(A = a) then
        perform pop action and move ahead the instruction pointer to
        point to next symbol
    else if (A is a terminal) then
```

```
        /*error occurred*/
    else if (M[A, a] is an error entry ) then
        /*error occurred*/
    else if (M [A, a] = A → B1 B2 B3...Bk then
    Begin
        output the production A → B1 B2 B3...Bk
        pop the stack
        push Bk, Bk-1, ..., B1 onto the stack, B1 on the top
    End
    set A to the top of stack
End
```

Figure 4.11 Non-recursive Predictive Parsing Algorithm

9. What are the advantages and disadvantages of table-driven predictive parsing?

Ans: Advantages:

- A table-driven parser can be easily generated from a given grammar. The parsing program is independent of the grammar but the parsing table depends on the grammar. With the use of FIRST and FOLLOW generation algorithms parsing table can be generated.
- Some entries in the parsing table have entries that point to the error recovery and reporting routines, which makes error recovery and reporting an easier task.

Disadvantage:

- Such type of parsers can work only on LL(1) grammars. Sometimes eliminination of left-factoring and left-recursion may not be sufficient to transform a grammar into LL(1) grammar.

10. Explain the error recovery strategies in predictive parsing.

Ans: A top-down predictive parser can be implemented by recursive-decent parsing or by table-driven parsing. The table-driven parsing predicts as to what terminals and non-terminals the parser expects from the rest of the input. An error can occur in the following situations:

- If the terminal on the top of a stack does not match the next input symbol.
- If a non-terminal `A` is on the top of stack, `x` is the next input symbol and the parsing table entry `M [A, x]` is empty.

Two commonly used error recovery schemes, namely, *panic mode recovery* and *phrase level recovery*.

Panic mode recovery is based on the idea that when an error occurs, the parser skips the input symbols until it finds a **synchronizing token** (a semicolon, }, or any other token with an unambiguous and clear role) in the input. A set of all synchronizing tokens is known as **synchronizing set**. The synchronizing set should be enough effective to recover from errors and it depends on the choice of synchronizing set. Some guidelines for constructing a synchronizing set are as follows:

- For a non-terminal `A`, place all the elements of `FOLLOW(A)` into the synchronizing set of `A`. Skip the tokens until an element of `FOLLOW(A)` is found and then pop `A` from the stack.
- For a non-terminal `A`, all the elements of `FIRST(A)` can also be added to the synchronizing set of `A`. This will help the parser in resuming the parsing according to `A`, if a symbol in `FIRST(A)` appears in the input.
- The production that derives $\in$ can be employed as a default production, if a non-terminal can produce the empty string. It may delay the error detection for some time but cannot cause an error to

be missed during error recovery. This approach is useful in reducing the number of non-terminals to be parsed.

- A terminal on the top of stack, which cannot be matched, is popped from the top of stack and a warning message indicating that the terminal was inserted is issued. After issuing the warning message, the parser can continue parsing as if the missing symbol is a part of the input. In effect, this approach takes all other tokens into the synchronizing set for a token.

Phrase level recovery is based on the idea of filling the blank entries in the predictive parsing table with pointers to error handling routines. The error handling routines can do the following:

- They can insert, modify, or delete any symbols in the input.
- They can also issue appropriate error messages.
- They can pop elements from the stack.

Pushing a new symbol onto the stack and alteration of the existing symbol in stack is problematic due to some reasons as given below:

- The steps performed by the parser may result in the derivation of a word that does not correspond to the derivation of any word in the language at all.
- There is a possibility of infinite loop formation during the alteration of stack symbol.

11. Explain bottom-up parsing with an example. Also discuss reduction in bottom-up parsing

Ans: Bottom-up parsing is a parsing method to construct a parse tree for an input string beginning at the leaves and growing towards the root. Bottom-up parsing can be considered as a process of reducing the input string to the start symbol of the grammar. At each reduction step, a particular substring matching the right hand side of a production is replaced by the non-terminal symbol on the left hand side of the production. Bottom-up parser can handle a large class of grammar.

For example, the steps to construct a parse tree of the token stream `a + b` with respect to a grammar G, `S → S * S|S + S|S - S|a|b` are shown in Figure 4.12.

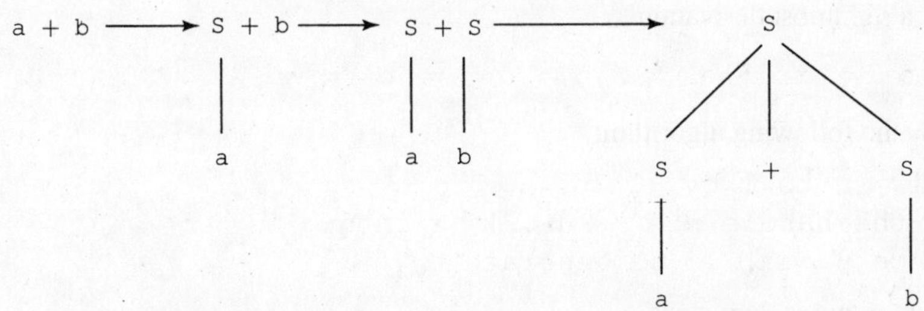

Figure 4.12 Steps to Construct a Bottom-up Parse Tree for `a + b`

The sequence starts with the input string `a + b`. The first reduction generates `S + b` by reducing `a` to `S`, using the production `S → a`. The second reduction generates `S + S` by reducing `b` to `S`, using the production `S → b`. Finally, the reduction of `S + S` to the start symbol `S` is done using the production `S → S + S`.

12. Define handle. Write a short note on handle pruning.

Ans: Handle of a string is a substring that matches the right hand side of some production and whose reduction to the non-terminal on the left hand side is a step along the reverse of some rightmost derivation. In other words, handle of a right sentential form γ is < $A \rightarrow \beta$, the location of β in γ >, such that replacing β by A at that position generates the previous right sentential form in a rightmost derivation of γ.

Formally, if

$$S \overset{*}{\underset{rm}{\Rightarrow}} \alpha A \gamma \underset{rm}{\Rightarrow} \alpha\beta\gamma$$

Then, $A \rightarrow \beta$ is a handle of $\alpha\beta\gamma$ at the location immediately after α.

Note that a certain sentential form may have many different handles but the right sentential forms of a non-ambiguous grammar have unique handle.

Consider the following productions to drive a string abbcde:

```
S → aABe
A → Abc|b
B → d
```

The rightmost derivation is as follows:

$$S \Rightarrow \underline{aABe} \Rightarrow aA\underline{d}e \Rightarrow a\underline{Abc}de \Rightarrow a\underline{b}bcde$$

It follows that

```
S → aABe is a handle of aABe in location 1.
B → d is a handle of aAde in location 3.
A → Abc is a handle of aAbcde in location 2.
A → b is a handle of abbcde in location 2.
```

Handle pruning: The process of discovering a handle and reducing it to the suitable left hand side is called handle pruning. A rightmost derivation in reverse (termed as **canonical reduction sequence**) is obtained by handle pruning. Handle pruning forms the basis of bottom-up parsing. Starting from a terminal string w, where w is a sentence of the grammar, let $w = \alpha n$, where αn is the nth right sentential form of some as yet unknown rightmost derivation.

To construct a rightmost derivation

$$S \Rightarrow \alpha_0 \Rightarrow \alpha_1 \Rightarrow \alpha_2 \Rightarrow \ldots \Rightarrow \alpha_{n-1} \Rightarrow \alpha_n \Rightarrow w$$

We can apply the following algorithm:

```
For i ← n to 1 by -1
Find the handle < A_i → β_i, k_i > in α_i
Replace β_i with A_i to generate α_(i-1)
```

This algorithm requires 2*n steps.

For example, the sequence of reductions to reduce abbcde to start symbol S is given in Table 4.2.

Table 4.2 Reduction of abbcde to S

Right Sentential Form	Handle	Reducing Production
a<u>b</u>bcde	b	A → b
a<u>Abc</u>de	Abc	A → Abc
aA<u>d</u>e	d	B → d
<u>aABe</u>	aABe	S → aABe
S		

13. Explain shift-reduce parsing with stack implementation.

Ans: Shift-reduce parsing is a kind of bottom-up parsing in which a stack is used to hold the grammar symbols and an input buffer is used to hold the remaining string to be parsed. The parser examines the input tokens and either shift (push) them onto a stack or reduce symbols at the top of the stack, replacing a right hand side by a left hand side. Though only shift and reduce symbols are considered as major operations but in fact a shift-reduce parser can make four actions:

- **Shift:** A shift action corresponds to pushing the next input symbol onto the top of stack.
- **Reduce:** A reduce action occurs when we have the right end of the handle at the top of the stack. To perform reduction, we locate the left end of the handle within the stack and choose a non-terminal on the left hand side of the corresponding rule to replace the handle.
- **Accept:** An accept action occurs when parser declares the successful completion of parsing.
- **Error:** An error action occurs when the parser finds a syntax error in the input and then parser calls an error recovery routine.

The symbol `$` is used to mark the bottom of the stack and the right end of the input string. Initially, the stack is empty and the input string is on the input buffer as shown below:

Stack	Input Buffer
`$`	`w$`

The parser performs a left-to-right scan through the input string to shift zero or more symbols onto the stack until it locates a prefix of the symbol (handle) on the top of the stack that matches the right hand side of a grammar rule. Then, the parser reduces the right hand side symbols on the top of the stack with the non-terminal occurring on the left hand side of the grammar rule. The parser repeats the process until it reports an error or a success message. The parsing is said to be successful, if the stack contains the start symbol and the input is empty as shown below:

Stack	Input Buffer
`$S`	`$`

Consider the following grammar:

`S → S + S|S * S|(S)|a`

To parse the input string `a + a`, a shift-reduce parser performs a sequence of steps as shown in Table 4.3.

Table 4.3 Shift-reduce Parsing Actions

Stack	Input Buffer	Action
`$`	`a + a$`	Shift
`$a`	`+a$`	Reduce `S → a`
`$S`	`+a$`	Shift
`$S+`	`a$`	Shift
`$S + a`	`a$`	Reduce s → a
`$S + S`	`$`	Reduce `S → S + S`
`$S`	`$`	Accept

14. Explain operator precedence parsing method of shift-reduce parsing.

Ans: Operator precedence parsing is a shift-reduce parsing technique that can be applied to operator grammar. An **operator grammar** is a small, but an important class of grammars in which no production rule can have:

- An ∈ production on the right hand side and
- Two adjacent non-terminals at the right hand side.

An operator precedence parser consists of the following:

- An input buffer containing the input string to be parsed,
- A stack containing the sequence of grammar symbols,
- An operator precedence relations table,
- A precedence parsing program,
- An output.

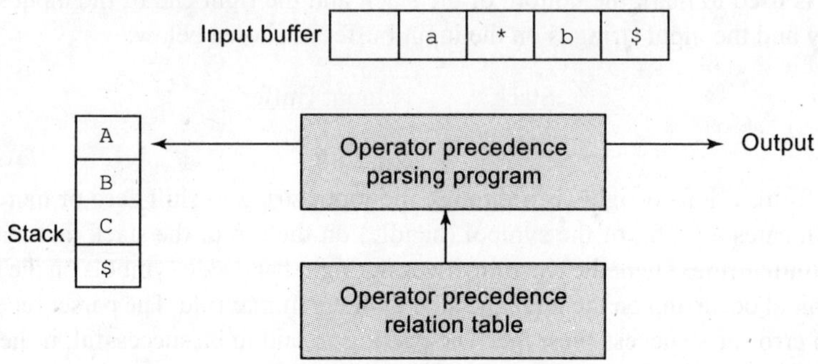

Figure 4.13 Operator Precedence Parser

There are three disjoint precedence relations that can exist between the pairs of terminals.

- a <· b   b has higher precedence than a.
- a ≐ b   b has same precedence as a.
- a ·> b   b has lower precedence than a.

Table 4.4 Operator Precedence Relations

	+	–	*	/	↑	id	(	)	$
+	·>	·>	<·	<·	<·	<·	<·	·>	·>
–	·>	·>	<·	<·	<·	<·	<·	·>	·>
*	·>	·>	·>	·>	<·	<·	<·	·>	·>
/	·>	·>	·>	·>	<·	<·	<·	·>	·>
↑	·>	·>	·>	·>	<·	<·	<·	·>	·>
id	·>	·>	·>	·>	·>			·>	·>
(	<·	<·	<·	<·	<·	<·	<·	≐	
)	·>	·>	·>	·>	·>			·>	·>
$	<·	<·	<·	<·	<·	<·	<·		

For example, consider the following grammar:

S → S * S|S * S|S - S|id

Then the input string id + id * id after inserting the precedence relations will be:

$ <· id ·> + <· id ·> * <· id ·> $

The operator precedence parsing algorithm to generate a parse tree for the input string is shown in Figure 4.14.

Input: An input string w$, a table holding precedence relations and the stack with initial symbol $.
Output: Parse tree.
Algorithm:

```
Set p to point to the first symbol of w$
Repeat forever
If ($ is on top of the stack and p points to $) then
   accept and exit the loop
else
Begin
   let terminal a is on the top of the stack and let
   b be the symbol pointed to by p
   If (a <· b or a ≐ b) then /* Shift */
   Begin
     push b onto the stack
     advance p to the next input symbol
   End
   else if (a ·> b) then /* Reduce */
     Repeat
       pop stack
     Until (the top of stack terminal is related by <·
           to the terminal most recently popped);
   else /*error occurred*/
End
```

Figure 4.14 Operator Precedence Algorithm

15. What are the advantages and disadvantages of operator precedence parsing?

Ans: Advantages:

- Operator precedence parsing is simple and easy to implement.
- Its parser is constructed by hand after knowing the grammar.
- Debugging is simple.

Disadvantages:

- Tokens like minus (−) are difficult to handle, as depending on whether it is being used as binary operator or unary operator it has two different values of precedence.
- It does take grammar as an input while generating a parser. This results in rewriting of the parser in case of any additions or deletions in the production rules, which is very cumbersome and time-consuming process.
- Only a small class of grammars like operator grammars can be parsed by this parsing technique.

16. Find FIRST of all the non-terminals of the following grammar:

S → ACB | CbB | Ba
A → da | BC
B → g | ∈
C → h | ∈

Ans:

1. FIRST (C) = FIRST (h) ∪ FIRST (∈)
   = {h} ∪ {∈}
   = {h, ∈}
2. FIRST (B) = FIRST (g) ∪ FIRST (∈)
   = {g} ∪ {∈}
   = {g, ∈}
3. FIRST (A) = FIRST (da) ∪ FIRST (BC)
   = {d} ∪ {FIRST (B) – {∈} ∪ FIRST (C)}
   = {d} ∪ {{g, ∈ } – {∈} ∪ {h, ∈} }
   = {d} ∪ {g} ∪ {h, ∈} [from (1) and (2)]
   = {d} ∪ {g, h, ∈}
   = {d, g, h, ∈}
4. FIRST(S) = FIRST (ABC) ∪ FIRST (CbB) ∪ FIRST (Ba)
   We first determine FIRST (ABC), FIRST (CbB), and FIRST (Ba) separately,

   FIRST (ABC) = FIRST (A) – {∈} ∪ FIRST (BC)
   = {d, g, h,∈} – {∈} ∪ First (B) – {∈} ∪ First (C)
   = {d, g, h} ∪ {g, ∈} – {∈} ∪ {h, ∈}
   = {d, g, h} ∪ {g} ∪ {h, ∈}
   = {d, g, h, ∈}

   FIRST (CbB) = FIRST (C) – {∈} ∪ FIRST (bB)
   = {h,∈} – {∈} ∪ {b}
   = {h} ∪ {b}
   = {h, b}

   FIRST (Ba) = FIRST (B) – {∈} ∪ FIRST (a)
   = {g, ∈} – {∈} ∪ {a}
   = {g} ∪ {a}
   = {g, a}

   Finally, by combining the value of FIRST (ABC), FIRST (CbB), and FIRST (Ba), we get
   FIRST (S) = {d, g, h,∈} ∪ {h, b} ∪ {g, a}
   = {d, g, h, b, a, ∈}

17. Consider the following grammar and show the handle of each right sentential form for the string (b, (b, b)).

E → (A) | b
A → A, E | E

Ans: The following sentential form will occur in reduction of `(b, (b, b))` to `S`.

1. `(b, (b, b))` (first `b` is the handle)
2. `(E, (b, b))` (`E` is the handle)
3. `(A, (b, b))` (first `b` is the handle)
4. `(A, (E, b))` (`E` is the handle again)
5. `(A, (A, b))` (`b` is the handle)
6. `(A, (A, E))` (`(A, E)` is the handle)
7. `(A, (A))` (`(A)` is the handle)
8. `(A, E)` (again `(A, E)` is the handle)
9. `(A)` (`(A)` is the handle)
10. `E` (finally string is reduced to starting non-terminal)

18. Consider the following grammar:

S → SAS
S → num
A → +
A → -
A → *
A → /

Explain why this grammar is not suitable to form the basis for a recursive-decent parser. Use left-factoring and left-recursion removal to obtain an equivalent grammar which can be used as the basis for recursive-decent parser.

Ans: Consider the following production:

S → SAS

If we put the value of S in place of first S at the right hand side in this production, the new production will be

S → SASAS

If we again put the value of S in place of first S at the right hand side, the new production will be

S → SASASAS

Thus, putting the value of S in place of first S at the right hand side again and again will result in an infinite loop. It shows that the given grammar suffers from the problem of left recursion. Hence, it cannot be the basis for recursive-decent parser.

If we put the value of A in the above production, we get

S → S + S
S → S - S
S → S * S
S → S/S
S → num

It results in the following production:

S → S + S|S - S|S * S|S/S|num

It still suffers from left recursion, which can be removed by following the algorithm discussed in Chapter 3. Now, we have the following productions:

```
S → num S'
S → +S S'|∈
S → -S S'|∈
S → *S S'|∈
S → /S S'|∈
```

This grammar does not suffer from left recursion, and hence, can form the basis for recursive-decent parser. The production will now become:

```
S → num S'|+S S'|-S S'|*S S'|/S S'|∈|num
```

19. Show that the given grammar is not LL(1).

E → iAcE|iAcEeE|a
A → b

Ans: Step 1: This grammar suffers from left factoring, so after removing left factoring

```
E → iAcEE'|a
E' → eE|∈
A → b
```

Step 2: Compute FIRST and FOLLOW of all non-terminals.

```
FIRST (E)    = {i, a}
FIRST (E')   = {e, ∈}
FIRST (A)    = {b}
FOLLOW (E)   = {$, e}
FOLLOW (E')  = {$, e}
FOLLOW (A)   = {c}
```

Now, to generate the parser table entries, follow these steps:

1. Considering the production E → iAcEE'
   FIRST (E) = FIRST (iAcEE') = {i}
   Since it does not contain any ∈.
   So, parse table [E, i] = E → iAcEE'
2. Considering the production E → a
   FIRST (E) = FIRST (a) = {a}
   Since it does not contain any ∈.
   So, parse table [E, a] = E → a
3. Considering the production E' → eE
   FIRST (E') = FIRST (eE) = {e}
   Since it does not contain any ∈.
   So, parse table [E', e] = E' → eE
4. Considering the production E' → ∈
   FIRST (E') = FIRST (∈) = {∈}
   Since it contains an ∈, we have to find out FOLLOW (E').
   FOLLOW (E') = {$, e}

So, parse table [E′, $] = E′ → ∈
Also, parse table [E′, e] = E′ → ∈

5. Considering the production A → b
FIRST (A) = FIRST (b) = {b}
Since it does not contain any ∈.
So, parse table [A, b] = A → b

The resultant parse table is shown in Table 4.5.

Table 4.5 LL(1) Parsing Table

	a	b	e	i	c	$
E	E → a			E → iAcEE′		
E′			E′ → eE E′ → ∈			E′ → ∈
A		A → b				

Multiple entries

The multiple entries in M [E′, e] field show that the the grammar is ambiguous and not LL(1).

Multiple-Choice Questions

1. Top-down parsing is a technique to find __________.
   (a) Leftmost derivation (b) Rightmost derivation
   (c) Leftmost derivation in reverse (d) Rightmost derivation in reverse
2. Predictive parsing is possible only for __________.
   (a) `LR(k)` grammar (b) `LALR(1)` grammar
   (c) `LL(k)` grammar (d) `CLR(1)` grammar
3. Which two functions are required to construct a parsing table in predictive parsing technique?
   (a) `CLOSURE()` and `GOTO()` (b) `FIRST()` and `FOLLOW()`
   (c) `ACTION()` and `GOTO()` (d) None of these
4. Non-recursive predictive parser contains __________.
   (a) An input buffer (b) A parsing table
   (c) An output stream (d) All of these
5. Which of these parsing techniques is a kind of bottom-up parsing?
   (a) Shift-reduce parsing (b) Reduce-reduce parsing
   (c) Predictive parsing (d) Recursive-decent parsing
6. Which of the following methods is used by the bottom-up parser to generate a parse tree?
   (a) Leftmost derivation (b) Rightmost derivation
   (c) Leftmost derivation in reverse (d) Rightmost derivation in reverse
7. Handle pruning forms the basis of __________.
   (a) Bottom-up parsing (b) Top-down parsing
   (c) Both (a) and (b) (d) None of these

8. In shift-reduce parsing, accept action occurs ___________.
   (a) When we have the right end of the handle at the top of the stack
   (b) When we have the left end of the handle at the top of the stack
   (c) When parser declares the successful completion of parsing
   (d) When the parser finds a syntax error in the input and calls an error recovery routine
9. Which of the following operators is hard to handle by the operator precedence parser?
   (a) Plus (+)
   (b) Minus (–)
   (c) Multiply (*)
   (d) Divide (/)
10. Given a grammar G:

```
T   →  BCTd | Bcd
CB  →  BC
Cc  →  cc
Bc  →  bc
Bb  →  b
```

    Which of the following sentences can be derived by G?
    (a) bcd
    (b) bbc
    (c) bcdd
    (d) bccd

Answers

1. (a) 2. (c) 3. (b) 4. (d) 5. (a) 6. (d) 7. (a) 8. (c) 9. (b) 10. (a)

5

LR Parsers

1. Explain LR parsers. What are its components?

Ans: LR parsers are efficient bottom-up parsers for a large class of context-free grammars. An LR parser is a non-backtracking shift-reduce parser in which 'L' indicates that they scan input from left to right and 'R' indicates that they construct a rightmost derivation in reverse. **LR parsing** is a method for syntactic recognition of programming languages. It makes use of tables to determine when a rule is complete and which additional tokens must be read from the source string. The term LR(k) can also be used to represent LR parser, where k indicates the number of input symbols of lookahead that are used in making parsing decision. Only the cases where the value of k is either 0 or 1 are of practical interest. If the value of k is not defined, it is taken as 1. So an LR parsing can be considered as an LR(1) parsing, that is, LR parsing with one symbol of lookahead.

Logically, an LR parser consists of two parts, a *driver routine* and a *parsing table*. The driver routine is same for all LR Parsers, only the parsing table changes from one parser to another. There are three major methods for constructing LR parsing table.

- **Simple LR or SLR parsing:** It is easy to implement but less powerful than other parsing methods.
- **Canonical LR or LR(1) parsing:** It is most general and powerful, but is tedious and costly to implement. It contains more number of states as compared to the SLR.
- **Lookahead LR or LALR(1) parsing:** It lies in between SLR and canonical LR in terms of power; however, it can be implemented efficient with little bit of effort. It contains the same number of states as the SLR.

The LR parser is a state machine. The architecture of an LR parser is shown in Figure 5.1. It consists of the following:

- an input buffer,
- a stack, containing a list of states $s_0\ s_1 \ldots s_m$, where s_m is on top,
- a goto table that suggests to which state it should move,
- an action table that provides a grammar rule to be applied,
- a set of CFL (context-free language) rules.

The LR parsing program uses the combination of the stack symbol on the top of the stack and the current input symbol to index the parsing table. The parsing table consists of two sub tables: ACTION table and GOTO table.

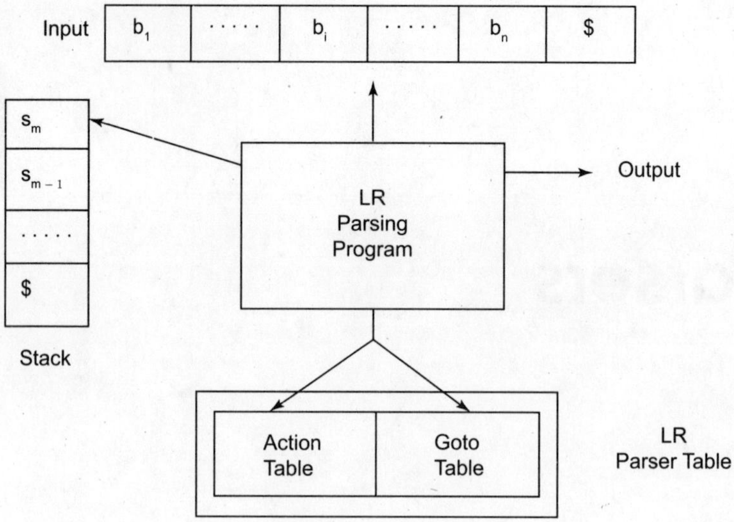

Figure 5.1 LR Parser

2. Why LR parsing is good and attractive? Also explain its demerits, if any.

Ans: LR parsing method is good and attractive due to the following reasons:

- LR parsing is the most common non-backtracking shift-reduce parsing.
- It is possible to construct the LR parsers for recognition of almost all programming language constructs for which CFG can be written.
- The class of grammars that can be parsed with predictive parsers can also be parsed using LR parsers. That is, the class of grammars that can be parsed with predictive parsers is a proper subset of those that can be parsed using LR parsers.
- An LR parser scans the input from left to right and while scanning it can detect the syntax errors as quickly as possible.

The main drawback of LR parsing is that for complex programming language grammars, the construction of LR parsing tables requires too much manual work. To reduce this manual work, we require an automated tool, known as **LR parser generator** that can generate an LR parser from a given grammar. Some available generators are YACC, bison, etc. These generators take context-free grammars as input and generate a parser for the input CFG. These generators also help in locating errors in the grammar, if any and generate error messages.

3. Explain ACTION and GOTO function in LR parsing.

Ans: While constructing a parsing table, we consider two types of functions: a parsing-action function `ACTION` and a goto function `GOTO`.

- **`ACTION` function:** The `ACTION` function takes a state s_m (the state on the top of stack) and a terminal b_i (the current input symbol) as input to take an action. The `ACTION` $[s_m, b_i]$ can have one of the four values:
  - **Shift `s`:** The action of the parser is to shift input `b` to the stack. Here, the parser uses state `s` to represent `b`.
  - **Reduce X $\rightarrow$ α:** The action of the parser is to reduce α on the top of the stack to head `X`.

- **Accept:** The parser accepts the input and announces successful parsing for the input string.
- **Error:** The parser finds an error and calls an error handling routine.

❑ **GOTO function:** The function GOTO can be defined as a set of states that takes a state and a grammar symbol as arguments and produces a new state. If GOTO [s_i, B] = s_j, then GOTO maps a state s_i and a non-terminal B to state s_j.

4. Explain configurations in LR parsing.

Ans: An LR parser configuration is a combination of two components. The first component is the stack content, which is a string of states and grammar symbols in the form $s_0\ X_1\ s_1\ X_2\ s_2 \ldots X_m\ s_m$, and the second component is the remaining input (the input which is still unexpanded). The state s_0 is the start state of the parser that does not represent a grammar symbol. It mainly serves as bottom-of-stack marker. The configuration of an LR parser can be represented as follows:

$$(s_0\ X_1\ s_1\ X_2\ s_2 \ldots\ldots X_m\ s_m,\ b_i\ b_{i+1} \ldots\ldots b_n\ \$)$$

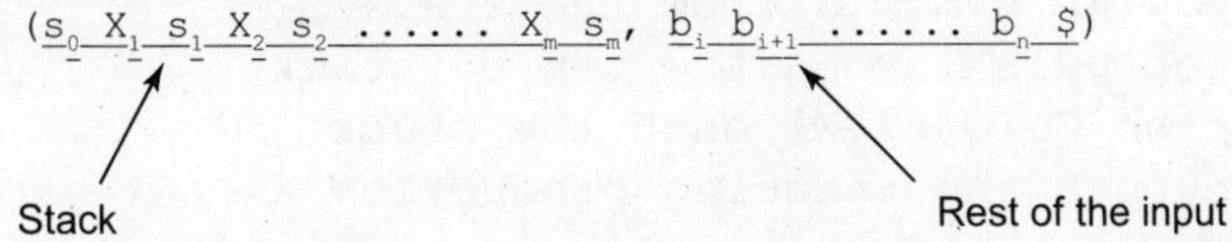

The combination of s_m (the state symbol on the top of the stack) and b_i (the current input symbol) decides the parser action by consulting the parsing action table. Initial stack contains only s_0. A configuration of an LR parsing represents the right sentential form

$$X_1 \ldots X_m,\ b_i\ b_{i+1} \ldots b_n\ \$$$

in the same way as that of shift-reduce parser. The only difference is that instead of grammar symbols, the stack contains those states from which grammar symbols can be recovered. That is, the grammar symbol X_j in right sentential form corresponds to state s_i in the configuration.

The configuration resulting after each of the four types of move is as follows:

❑ If ACTION[s_m, b_i] = shift s, the parser performs shift operation, that is, it shifts the s state (next state) onto the stack. The configuration now becomes:

$$(s_0\ s_1 \ldots s_m\ s,\ b_{i+1} \ldots b_n \$)$$

Note that the current input symbol is b_{i+1} and there is no need to hold the symbol b_i on the stack, as it can be recovered from S if required.

❑ If ACTION[s_m, b_i] = reduce X $\rightarrow$ α, the parser performs a reduce operation. The new configuration is:

$$(s_0\ s_1 \ldots s_{m-p}\ s,\ b_i\ b_{i+1} \ldots b_n \$)$$

where

p is the length of α (the body of the reducing production)

s = GOTO [s_{m-p}, X]

The parser first pops the p state symbols from the stack, which exposes the state s_{m-p}. Then, it pushes the state s, which is the entry for GOTO [s_{m-p}, X], onto the stack. Note that b_i is still the current input symbol, that is, the reduce operation does not alter the current input symbol.

❑ If ACTION[s_m, b_i] = accept, it indicates the completion of parsing and the string is accepted.

❑ If ACTION[s_m, b_i] = error, an error is encountered by the parser and an error recovery routine is called.

5. Write the LR parsing algorithm.

Ans: Let us consider an input string w and an LR parsing table for grammar G with ACTION and GOTO functions. The algorithm for LR parsing is given in Figure 5.2.

```
Set instruction pointer to point to first symbol b of w$
Do
    Set s as the state on top of the stack
       If(ACTION[s, b] = shift p) then
          push p onto the stack and move ahead instruction
          pointer to point to next symbol
       else if(ACTION[s, b] = reduce X → α) then
       Begin
          pop |α| symbols from the stack
          let p is now on the top of stack
          push GOTO[p, X] onto the stack
          output the reducing production X → α
       End
       else if(ACTION[s, b] = accept) then
          exit the loop /* parsing is completed */
       else
          /* error occurred */
While(1)
```

Figure 5.2 LR Parsing Algorithm

6. Define LR(0) items and LR(0) automaton. Also explain the canonical LR(0) collection of items.

Ans: An **LR(0) item** (in short **item**) is defined as a production of grammar G having a dot at some position of the right hand side of the production. Formally, an item describes how much of a production has already been seen on the input at some point in the parsing process. For example, consider the following four items created by the production X → ABC:

- X → .ABC, which indicates that a string derivable from ABC is expected next on the input.
- X → A.BC, which indicates that a string derivable from A has already been seen on the input and now a string derivable from BC is expected.
- X → AB.C, which indicates that a string derivable from AB has already been seen and now a string derivable from C is expected on the input.
- X → ABC., which indicates that the body of the production has already been seen, and now it is time to reduce it to X.

The **canonical LR(0) collection** is a collection of sets of LR(0) items, which provides the basis for constructing a DFA that is used to make parsing decisions. Such an automaton is called an **LR(0) automaton**. The states in the LR(0) automaton correspond to the sets of items in the canonical LR(0) collection. The canonical LR(0) collection for a grammar can be constructed by defining its augmented grammar and two functions, CLOSURE and GOTO.

For a grammar G with start symbol S, G′ will be the augmented grammar of G, with a new start symbol S′ and production S′ → S. This new generated production indicates that when the parser should stop parsing and announce the acceptance of the input. Thus, we can say that acceptance occurs only when the parser is about to reduce by S′ → S.

- **CLOSURE:** Let I be the set of items for a grammar G, then we construct a set of items CLOSURE(I) from I by following these steps:
  - Add every item in I to CLOSURE(I).
  - If A → α.Bβ is in CLOSURE(I), where B is a non-terminal and B → γ is a production in G, then add the item B → .γ to CLOSURE(I), if it is not already present in it.
  - Repeat step 2 until there are no more items to be added in CLOSURE(I).

  In step 2, A → α.Bβ in CLOSURE(I) represents that a substring derivable from Bβ is expected to be seen as input at some point in the parsing process. The substring derivable from Bβ will have the prefix derivable from B by applying one of the productions of B. Thus, we include items for all productions of B in CLOSURE(I). For this reason, we include B ↦ .γ in CLOSURE(I).
- **GOTO:** If I is a set of items, X is a grammar symbol, and an item (A → α.Xβ) is in I, then the function GOTO(I, X) is defined as the closure of the set of all items (A → αX.β). The GOTO function basically defines the transitions in the LR(0) automaton. The states in LR(0) automaton are represented as set of items, and GOTO(I, X) defines the transition from state for I for the given input X.

7. What is a Simple LR parser or SLR parser?

Ans: **SLR parser** is the simplest LR parser technique generating the parsing tables like LR(0) Parser. But unlike LR(0) parser, it only performs a reduction with a grammar rule A → w if the next input symbol is in FOLLOW(A). This parser can prevent shift-reduce and reduce-reduce conflicts occurring in LR(0) parsers. Therefore, it can deal with more grammars. A grammar that can be parsed by an SLR parser is called an **SLR grammar**. For example, a grammar that can be parsed by SLR parser but not by an LR(0) is given below:

E → 1|E
E → 1

8. Explain how to construct an LR(0) parser.

Ans: The various steps for constructing an LR(0) parser are given below:

1. For a grammar G with a start symbol S, construct an augmented grammar G′ with a new start symbol S′ and production S′→ S.
2. Compute the canonical collection of LR(0) items of grammar G.
3. Find the state transitions for state i for all non-terminals X using the following GOTO function:

   I_1 = GOTO(I_0, X).
4. Construct new states with the help of CLOSURE(I) and GOTO(I, X) functions, and for each state construct new LR(0) items.
5. Repeat step 4 until there are no more transitions left on input for which state transitions can be constructed.
6. With all computed states I_0, I_1, . . . , I_n, construct the transition graph by keeping LR(0) items of each I_i in single node and linking these nodes with suitable transitions evaluated with GOTO.
7. Constitute a parse table using SLR table construction algorithm.
8. Apply the shift-reduce action to verify whether the input string is accepted or any conflict has occurred.

9. Discuss the algorithm to construct an SLR parsing table.

Or

How to make an ACTION and GOTO entry in SLR parsing table?

Ans: SLR parsing tables are constructed using two functions, namely, ACTION and GOTO. It begins with LR(0) items and LR(0) automaton, that is, for a grammar G, an augmented grammar G′ with a new start symbol S′ is constructed. For this augmented grammar G′, the canonical collection of items C is computed with the help of GOTO function. And then we construct the ACTION and GOTO entries in the parsing table using the following algorithm:

Step 1: For any augmented grammar G′, construct the collection of sets of LR(0) items $C = \{I_0, I_1, \ldots, I_n\}$.

Step 2: For each state in C, construct a row in SLR table and name the rows from 0 to n. Partition the columns into ACTION and GOTO, where ACTION will have all terminal symbols of grammar G along with symbol $, and GOTO will have all the non-terminals of G.

Step 3: For each I_i construct a state i. The action entries for state i in SLR parsing table are determined using the following rules:

- If $[A \rightarrow \alpha.b\beta]$ is in I_i and $GOTO(I_i, b) = I_j$, where b is a terminal, then ACTION[i, b] = "shift j", which implies an entry corresponding to state i and terminal j is made in the ACTION part.
- For all b in FOLLOW(A), if $[A \rightarrow \alpha.]$ is in I_i, then ACTION[i, b] = "reduce $A \rightarrow \alpha$".
- If $[S' \rightarrow S.]$ is in I_i, then ACTION[i, $] = "accept".

If any conflicting actions occur from these rules, we will not consider the grammar as SLR(1). In that case, this algorithm will not be able to produce a valid parser.

Step 4: The goto entries for state i in the SLR parsing table can be determined using the following rule:

If $GOTO(I_i, A) = I_j$, then GOTO[i, A] = j.

Step 5: The undefined entries are made 'error' entries.

Step 6: The initial state of parser is one constructed from the set of items containing $[S' \rightarrow .S]$.

The parsing table containing ACTION and GOTO functions defined by the above algorithm is known as **SLR(1) parsing table** for G, and the LR parser that makes use of this table is known as **SLR(1) parser** for G. Here, 1 in SLR(1) denotes the single lookahead symbol.

10. What are the demerits of SLR parser?

Ans: The SLR parser has certain demerits given below:

- A state may include a final item and a non-final item. This may result in a shift-reduce conflict.
- A state may include two different final items. This might result in reduce-reduce conflict.

11. Explain viable prefixes.

Ans: The stack in a shift-reduce parser for a grammar G can hold the prefixes of right-sentential form. However, not all the prefixes of right sentential forms can appear in the stack, that is, the stack is allowed to hold only some of the prefixes. These prefixes that the stack can hold are known as **viable prefixes**. A viable prefix is so called because it is always possible to add terminal symbols to the end of a viable prefix to obtain a right-sentential form. The fact that the LR(0) automata recognize viable prefixes forms the basis for SLR parsing. For example, we consider the following item:

$$X \rightarrow \beta_1.\beta_2$$

It is valid for a viable prefix $\alpha\beta_1$ if there exists a derivation

$$S' \underset{rm}{\overset{*}{\Rightarrow}} \alpha Xw \underset{rm}{\Rightarrow} \alpha\beta_1\beta_2 w$$

We can deduce the information about whether to take shift action or reduce action from the fact that $X \rightarrow \beta_1.\beta_2$ is valid for $\alpha\beta_1$ as follows:

- If $\beta_2 \neq \in$, it indicates that we have not yet shifted the handle onto the stack, so need to perform a shift action.
- If $\beta_2 = \in$, it indicates as if $X \rightarrow \beta_1$ is the handle, and we should reduce by this production.

Thus, it is clear that for the same viable prefix, two valid items may indicate to do different things. Such conflicts can be resolved by looking ahead the next input symbol.

Generally, an item can be valid for many viable prefixes. The set of items valid for a viable prefix γ can be computed by determining the set of items that can be reached from the initial state along the path labeled γ in the LR(0) automaton for the grammar.

12. What is the canonical LR parser?

Ans: A canonical LR (CLR) parser is more powerful than LR parser as it makes full use of one or more lookahead symbols. It contains some extra information in the form of a terminal symbol, as a second component in each item of state. Thus, in CLR parser, an item can be described as follows:

$$[A \rightarrow \alpha.\beta,\ a]$$

where

$A \rightarrow \alpha\beta$ is a production, and
a is the terminal symbol or right end marker \$.

Such an item is defined as **LR(1) item**, where 1 refers to the length of second component, called the **lookahead** of the item. If $\beta \neq \in$, then the lookahead does not effect the item $[A \rightarrow \alpha.\beta,\ a]$. However, if the item has the form $[A \rightarrow \alpha.,\ a]$, then it calls for a reduction by $A \rightarrow \alpha$ only if the next input symbol is a. That is, we are compelled to reduce $A \rightarrow \alpha$ only on those input symbols a for which $[A \rightarrow \alpha.,\ a]$ is an LR(1) item in the state on the top of the stack.

13. Write the algorithm for computation of sets of LR(1) items.

Or

Define CLOSURE(I) and GOTO(I, X) functions for LR(1) grammar.

Ans: The algorithm for computing the sets of LR(1) items is basically the same as that of the canonical sets of LR(0) items—only the procedures for computing the CLOSURE and GOTO need to be modified as shown in Figure 5.3.

In Figure 5.3, the function items() is the main function that calls the CLOSURE and GOTO functions for constructing the sets of LR(1) items for grammar G'.

```
procedure CLOSURE(I)
Begin
     Do
          For (each item [A → α.Bβ, a] in I)
          For (each production B → γ in G')
               For (each terminal b in FIRST(βa))
```

```
            add [B → . γ, b] to I
    While there are some items to be added to set I
    return I
End

procedure GOTO(I, X)
Begin
    Initialize J to be the empty set
    For (each item [A → α.Xβ, a] in I)
        add item [A → α.Xβ, a] to set J
    return CLOSURE(J)
End

void items(G')
Begin
    C = CLOSURE([S' → .S, $])
    Do
        For (each set of items I in C)
            For (each grammar symbol X)
                If (GOTO(I, X) is not empty and not in C)
                    add GOTO(I, X) to C
    While there are some sets of items to be added to C
End
```

Figure 5.3 Computation of Sets of LR(1) Items for Grammar G′

14. Give the algorithm for the construction of canonical LR parsing table.

Ans: Canonical LR parsing tables are constructed by the LR(1) ACTION and GOTO functions from the set of LR(1) items. The ACTION and GOTO entries are constructed in the parsing table using the following algorithm:

Step 1: For any augmented grammar G′, construct the collection of sets of LR(0) items $C' = \{I_0, I_1, ..., I_n\}$.

Step 2: For each state in C, construct a row in CLR table and name the rows from 0 to n. Partition the columns into ACTION and GOTO, where ACTION will have all terminal symbols of grammar G along with symbol $, and GOTO will have all the non-terminals of G.

Step 3: For each I_i construct a state i. The action entries for state i in CLR parsing table are determined using the following rules:

- If [A → α.aβ, b] is in I_i and GOTO(I_i, a) = I_j, where a is a terminal, then ACTION[i, a] = "shift j".
- If [A → α., a] is in I_i, and A ≠ S′ then ACTION[i, a] = "reduce A → α.".
- If [S′ → S., $] is in I_i, then ACTION[i, $] = "accept".

If any conflicting actions occur from the above rules, we will not consider the grammar as LR(1). In that case, this algorithm will not be able to produce a valid parser.

Step 4: The goto entries for state i in the CLR parsing table can be determined using the following rule:

If GOTO(I_i, A) = I_j, then GOTO[i, A] = j.

Step 5: The undefined entries are made 'error' entries.

Step 6: The state of parser that is constructed from the set of items containing `[S' → .S, $]` is the initial state of the parser.

The table containing the parsing action and goto functions produced by the canonical LR parsing table algorithm is called the **canonical LR(1) parsing table**, and the LR parser that uses this table is called a **canonical LR(1) parser**. If there are no multiple defined entries in the parsing action function, then the given grammar is called an LR(1) grammar.

15. What is LALR parsing? Give the algorithm for the construction of LALR parsing table.

Ans: **LALR** parsers or **lookahead LR** parsers are specialized form of LR parsers that work on most of the programming language and can be implemented more efficiently. LALR parsers lie in between SLR parsers and canonical LR parsers in terms of power of parsing grammars. That is, they deal with more grammars than that of SLR parsers but less than that of LR(1) parsers. LALR parsing technique is most frequently used because the tables generated by it are considerably smaller than the canonical LR tables. Some of the most commonly used parser generators programs such as YACC and bison use the LALR parsing technique for constructing the parsing table.

The LALR parsing table can be constructed from the collection of merged set of items. The main principle behind the construction of LALR parsing table is to merge the set of LR(1) items having the same core to form the condensed set of LR(1) items. The sets containing the same first components of the LR(1) items are said to have the same **core**. That is, the components appearing at the left hand side of the dot symbol in a production are same; only the lookahead symbol(s) (the symbol appearing at the right hand side of the dot symbol) varies. Thus, the number of sets of items produced by LALR method is smaller than the number of sets produced by the CLR method. The algorithm for the construction of LALR parsing table is shown below:

Step 1: For any augmented grammar G', construct the collection of sets of LR(1) items $C = \{I_0, I_1, I_2, \ldots, I_n\}$.

Step 2: For each core present among the sets of LR(1) items, determine the sets that have that core and replace them with their union. Suppose the resulting sets of LR(1) items be $C' = \{J_0, J_1, J_2, \ldots, J_k\}$, where J_i is a union of one or more sets of LR(1) items having the same core. Mathematically, $J_i = \{I_1 \cup I_2 \cup I_3 \cup \ldots \cup I_m\}$.

Step 3: For each J_i construct a state i. The action entries for state `i` in LALR parsing table are determined in the same way as that of canonical LR parser. If any conflicting actions occur, we will not consider the grammar as LALR(1). In that case, this algorithm will not be able to produce a valid parser.

Step 4: The goto table is constructed by taking the union of all sets of items having common core. Since $I_1, I_2, \ldots, I_m$ all have the common core, the cores of $GOTO(I_1, X), GOTO(I_2, X), \ldots, GOTO(I_m, X)$ will also be same. If we consider U as the union of all sets of items having the common core as $GOTO(I_1, X)$, then $GOTO(J, X) = U$, where $J = I_1 \cup I_2 \cup \ldots \cup I_m$.

16. Specify the merits and demerits of LALR parser.

Ans: The merits of LALR parser are as follows:

- The LALR parser tables are smaller than the CLR tables.
- An LALR grammar can easily express the syntactic structure of programming languages.
- The LALR parsing technique can be implemented more efficiently.

- The LALR parser provides a good trade off between power and efficiency.
- Merging of items never introduces shift-reduce conflict unless the conflict is already present in LR(1) configuration sets.

The demerits of LALR parser are as follows:

- The construction of parser table from the collection of LR(1) items requires too much space and time.
- Merging of items may introduce reduce-reduce conflict. In case reduce-reduce conflict arises, the grammar is not considered as LALR(1).

17. Differentiate between SLR and LALR.

Or

Why LALR parser is considered over SLR?

Ans: In LALR parsing, the reduce entries are made using lookahead sets whereas in SLR, reduce entries are made using succeed sets. The **lookahead set** for LR(0) item `I` consists of only those symbols that are expected to be appeared after `I`'s right hand side has been parsed. On the other hand, the **succeed set** consists of all those symbols that are supposed to appear after `I`'s left hand side non-terminal. The lookahead set is more specific to parsing context and provides a finer distinction than the succeed set. In SLR parsing, shift-reduce conflict arises whereas merging of items does not introduce any shift-reduce conflict in LALR parsing. Reduce-reduce conflict may occur in LALR parsing.

18. Discuss how YACC can be used to generate a parser?

Ans: YACC stands for **yet another compiler-compiler**. It is an LALR parser generator which is basically available as a command on UNIX system. The first version of YACC was created by S.C. Johnson in early 1970s. It is a tool that compiles a source program and translates it into a C program that implements the parser.

For example, consider a file `translate.y`. The YACC compiler converts this file into a C program `y.tab.c` using the LALR algorithm. The program `y.tab.c` is basically a representation of LALR parser written in C language. This program is then compiled along with the `ly` library to generate the object code `a.out`, which then performs the translation specified by the original YACC program. An input/output translator constructed using YACC is given in Figure 5.4.

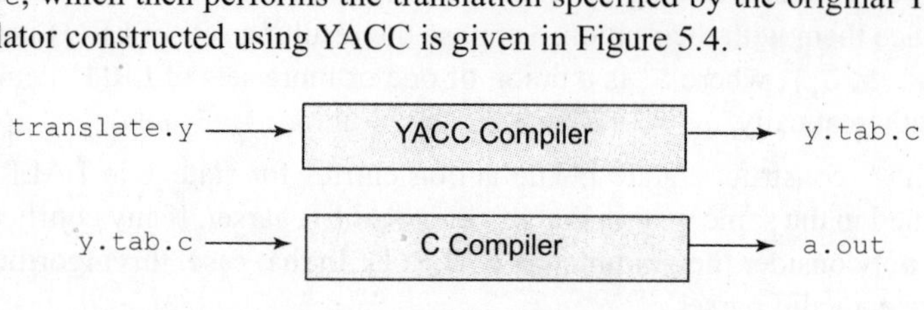

Figure 5.4 Input/Output Translator with YACC

A YACC source program consists of the following three parts:

```
declaration
%%
transition rules
%%
supporting C routines
```

- **Declaration section:** This section consists of two optional sections. The first section contains ordinary C declarations delimited by `%{` and `%}`. For example, it may contain `#include` preprocessors as given below:

```
%{
#include<conio.h>
%}
```

The second section contains the declarations of grammar tokens. For example, the following statement declares `DIGIT` as a token:

```
%token DIGIT
```

- **Translation rules section:** This section includes the grammar productions along with their semantic actions. For example, the productions of the following form:

```
head → alternative1 |alternative2 | . . . | alternativen
```

can be written in YACC as follows:

```
head :   alternative1   {semantic action 1}
     |   alternative2   {semantic action 2}
     . . .
     |   alternativen   {semantic action n}
     ;
```

The semantic action includes the values associated with the non-terminals of the head. The symbol `$$` is used to refer these values. For example, consider the productions,

```
E → E * F | F
```

The YACC specification for the above productions can be written as follows:

```
expr  : expr '*' factor   {$$ = $4 * $5;}
      | factor
      ;
```

- **Supporting C-routines section:** This section includes the lexical analyzer `yylex()` that produces pairs consisting of token name and their associated values. Th attribute values are communicated to the parser through the variable `yylval` already defined in YACC.

19. Explain ambiguity in LR parsers and the ways to eliminate it.

Ans: LR parser involves two conflicts, namely, *shift-reduce conflict* and *reduce-reduce conflict*.

- **Shift-reduce conflict:** This conflict occurs when it is difficult to decide whether to shift a token or to reduce a handle. For example, consider the dangling-else grammar:

```
statement   → if condition then statement|
              if condition then statement else statement|
              other statement
```

Suppose the status of the stack and next input symbol at some point is as follows:

Stack	Input
...**if** condition **then** statement	**else**...$

Depending on what follows the else on the input, the parser may choose to *reduce* **if** condition **then** statement to statement, or it may choose to *shift* else and then look for another statement to complete the alternative **if** condition **then** statement **else** statement. This gives rise to shift-reduce conflict since the parser cannot decide whether to shift **else** onto the stack or to reduce **if** condition **then** statement.

This ambiguity can be resolved by matching each **else** with its just preceding unmatched **then**. Thus, in our case the next action would be shift **else** onto the stack because it is associated with the previous **then**.

- **Reduce-reduce conflict:** This conflict occurs when we know we have a handle but the next input symbol and the stack's contents are not enough to determine which production is to be used in a reduction. For example, consider a language in which a procedure can be invoked by giving the procedure name along with the parameters surrounded by parentheses, and array references are also made using the same syntax. Some of the productions for the grammar of our language would be as follows:

```
statement → id(parameter_list)                          (1)
statement → expression: = expression                    (2)
parameter_list → parameter_list, parameter              (3)
parameter_list → paramater                              (4)
parameter → id                                          (5)
expression → id(expression_list)                        (6)
expression_list → expression_list, expression           (7)
expression_list → expression                            (8)
expression → id                                         (9)
```

Let us consider an input string A(X, Y). The token stream for the given input string for the parser is id(id, id). Now, the configuration of the parser after shifting the initial three tokens onto the stack is as follows:

Stack	Input
...**id(id**	**,id)**...$

It is clear that we need to reduce the id that is on top of the stack, but it is not clear that which production needs to be used for reduction. If A is a procedure name then production (5) needs to be used, and if A is an array then production (9) needs to be followed. Thus, reduce-reduce conflict occurs.

This conflict can be resolved by changing the token id in production (1) by procid, and by using a more sophisticated lexical analyzer that returns token procid for an identifier which is a procedure name, and id for an array name. Before returning a token, the lexical analyzer needs to consult the symbol table. Now, if A is a procedure then after this modification of token stream the configuration of the parser would be:

Stack	Input
...**procid(id**	**,id)**...$

20. Explain error recovery in LR parsing.

Ans: An LR parser detects an error when it finds undefined entries in the LR parsing table or when an empty entry for an input combination is made in the table. Checking of `goto` tables never results in errors. The requirement of error recovery is to enable the parser to detect syntax errors and reporting them to the user for correction. There are two error recovery strategies in LR parsing, namely, *panic mode* and *phrase level error recovery*.

- **Panic mode error recovery:** The panic mode error recovery can be implemented in LR parsing in the following manner:
  - The stack is scanned down until a state `s` having a `goto` on a particular non-terminal `X` is found. The non-terminal would be an expression or statement.
  - Then we discard zero or more input symbols until we found a symbol `x` that follows `X`. If `X` is a statement, then `x` would be a semicolon or end.
  - Parser then pushes the state `GOTO[s, X]` onto the stack and resumes normal parsing.
- **Phrase level error recovery:** The phrase level error recovery can be implemented in LR parsing in the following manner:
  - Each error entry in the LR parsing table is examined and on the basis of language usage, the most likely programmer error that would result in that error is determined.
  - An appropriate error recovery procedure is then constructed. deemed
  - For each error entry, the top of the stack or the first input symbols, or both are then modified. For example, a comma is replaced by a semicolon or a missing semicolon is inserted or an extraneous semicolon is deleted.

21. Construct an LR(0) parsing table for the following grammar G:

```
P → Q )
P → Q, R | ( R, R
R → {num, num}
```

Ans: The augmented grammar `G'` for the above grammar `G` is:

```
P' → P                (rule 0)
P  → Q )              (rule 1)
P  → Q, R             (rule 2)
P  → ( R, R           (rule 3)
R  → {num, num}       (rule 4)
```

Item set number 0, I_0

```
 P' → .P
+P  → .Q )
+P  → .Q, R
+P  → .( R, R
```

where '+' denotes the closure of the item `P' → .P` and it is not any terminal.
In I_0 symbols just after the dot are `P`, `Q`, and `(`.

Item set number 1, I_1 (for the symbol P of I_0)

```
P' → P.
```

Item set number 2, I_2 (for the symbol Q of I_0)

```
P → Q . )
P → Q ., R
```

Item set number 3, I_3 (for the symbol (of I_0)

```
P → ( . R, R
+R → . {num, num}
```

In I_1, no symbol is left after the dot.
In I_2, symbols just after the dot are) and , .

Item set number 4, I_4 (for the symbol) of I_2)

```
P → Q ) .
```

Item set number 5, I_5 (for the symbol , of I_2)

```
P → Q, . R
+R → . {num, num}
```

In I_3, symbols just after the dot are R and {.

Item set number 6, I_6 (for the symbol R of I_3)

```
P → ( R . , R
```

Item set number 7, I_7 (for the symbol { of I_3)

```
R → {. num, num}
```

In I_4, no symbol is left after the dot.
In I_5, symbols just after the dot are R and {.

Item set number 8, I_8 (for the symbol R of I_5)

```
P → Q, R .
```

Whereas, symbol { of I_5 is already processed in I_7.
In I_6, symbol just after the dot is , .

Item set number 9, I_9 (for the symbol , of I_6)

```
P → ( R, . R
+R → .{num, num}
```

In I_7, symbol just after the dot is num.

Item set number 10, I_{10} (for the symbol num in I_7)

```
R → {num . , num}
```

In I_8, no symbol is left after the dot.
In I_9, symbols just after the dot are R and {.

Item set number 11, I_{11} (for the symbol R of I_9)

```
P → ( R, R .
```

Also symbol `{` of I_9 is already processed in I_7.
In I_{10}, symbol after the dot is `,`.

Item set number 12, I_{12} (for the symbol `,` of I_{10})

```
R → {num,  . num}
```

In I_{11}, no symbol is left after the dot.
In I_{12}, symbol after the dot is `num`.

Item set number 13, I_{13} (for the symbol N of I_{12})

```
R → {num, num . }
```

In I_{13}, symbol after the dot is `}`.

Item set number 14, I_{14} (for the symbol } of I_{13})

```
R → {num, num}.
```

Now, in I_{14}, no symbol is left after the dot.
Thus, the transition table becomes:

Item Set	)	,	(	{	Num	}	P	Q	R
0			3				1	2	
1									
2	4	5							
3				7					6
4									
5				7					8
6		9							
7					10				
8									
9				7					11
10		12							
11									
12					13				
13						14			
14									

Thus, the action/goto table becomes:

	Action							Goto		
Item Set	)	,	(	{	Num	}	$	P	Q	R
0			S_3					1	2	
1							acc			
2	S_4	S_5								
3				S_7						6

(Continued)

(Continued)

	Action							Goto		
Item Set	**)**	**'**	**(**	**{**	**Num**	**}**	**$**	**P**	**Q**	**R**
4	r_1	r_1	r_1	r_1	r_1	r_1	r_1			
5				S_7						8
6		S_9								
7					S_{10}					
8	r_2	r_2	r_2	r_2	r_2	r_2	r_2			
9				S_7						11
10		S_{12}								
11	r_3	r_3	r_3	r_3	r_3	r_3	r_3			
12					S_{13}					
13						S_{14}				
14	r_4	r_4	r_4	r_4	r_4	r_4	r_4			

This is the resultant LR(0) parsing table.

22. Construct the sets of LR(0) items for the following grammar:

E → E + E | E * E | (E) | id

And also construct the parsing table.

Ans: The given grammar is ambiguous as the precedence or associativity of the operator * and + is not specified. For the given grammar, the augmented grammar is as follows:

```
E' → E
E  → E + E
E  → E * E
E  → (E)
E  → id
```

The set of LR(0) items is as follows:

```
I0: E' → .E
    E  → .E + E
    E  → .E * E
    E  → .(E)
    E  → .id

I1: E' → E
    E  → E. + E
    E  → E. * E

I2: E  → (.E)
    E  → .E + E
    E  → .E * E
    E  → (E)
    E  → .id

I3: E  → id.
```

```
I4:  E  →   E + .E
     E  →  .E + E
     E  →  .E * E
     E  →  .(E)
     E  →  .id

I5:  E  →   E * .E
     E  →  .E + E
     E  →  .E * E
     E  →  .(E)
     E  →  .id

I6:  E  →  (E.)
     E  →  E. + E
     E  →  E. * E

I7:  E  →  E + E.
     E  →  E. + E
     E  →  E. * E

I8:  E  →  E * E.
     E  →  E. + E
     E  →  E. * E

I9:  E  →  (E).
```

Now, the parsing table for the above set of LR(0) items will be:

State	Action						Goto
	id	+	*	(	)	$	E
0	S_3			S_2			1
1		S_4	S_5			acc	
2	S_3			S_2			6
3		r_4	r_4		r_4	r_4	
4	S_3			S_2			8
5	S_3			S_2			8
6		S_4	S_5		S_9		
7		r_1	S_5		r_1	r_1	
8		r_2	r_2		r_2	r_2	
9		r_3	r_3		r_3	r_3	

23. Every SLR(1) is unambiguous but there are few unambiguous grammars that are not SLR(1). Verify this for the following productions.

```
S → L = R
S → R
L → * R
L → id
R → L
```

Ans: The augmented grammar G′ for the above productions is as follows:

```
S' →  S
S  →  L = R
S  →  R
L  →  *R
L  →  id
R  →  L
```

The canonical collection of LR(0) items for G are as follows:
Starting with closure (S' → .S), we get I_0

```
S' →  .S
S  →  .L = R        by Rule 1 of closure, A → α . Bβ
S  →  .R            B → .γ
L  →  .* R          closure(1), by Rule 3, B → . γ
L  →  .id           B → .γ in closure(2)
R  →  .L            B → .γ closure(2)

I1 = GOTO (I0, S)
   = Closure (S' → S.), we obtain
   S'→ S.

I2 = GOTO (I0, L)
   = Closure (S → L .= R) ∪ Closure (R → L.), we obtain
   S →                 L. = R
   R →                 L.

I3 = GOTO (I0, R)
   = Closure (S → R.), we obtain
   S → R.

I4 = GOTO (I0, *)
   = Closure (L → *.R), we obtain
   L → *.R
   R → .L              B → .γ in closure (L → *.R)
   L → .*R             B → .γ in closure (R → .L)
   L → .id             B → .γ in closure (L → .id)

I5 = GOTO (I0, id)
   = Closure(B → id.), we obtain
   L → id.

I6 = GOTO (I2, =)
   = Closure (S → L = .R), we obtain
   S → L = .R
   R → .L              B → .γ in closure (S → L = .R)
   L → .* R            B → .γ in closure (R →.L)
   L → .id             B → .γ in closure (R → .L)

I7 = GOTO (I4, R)
   = Closure (L → *R.), we obtain
   L → *R.
```

```
I8 = GOTO (I4, L)
   = Closure (R → L.), we obtain
   R → L.
I9 = GOTO (I6, R)
   = Closure (S → L = R.), we obtain
   S → L = R.
```

Thus, we get the canonical LR(0) items, now to verify whether this grammar is SLR(1) or not by applying rule (3) of SLR parsing table algorithm.

- Consider the production, S → L. = C in I_2.
- Compare it with A → α. aβ, we obtain that a is =.
- We know that GOTO(I_2, =) = I_6; therefore, by applying rule 3(a) of SLR parser table algorithm, we obtain ACTION(2, =) = S_6.
- But one more production exists in I_2, that is, R → L. By applying rule 3(b) of SLR table algorithm, and comparing R → L. with A → α. and FOLLOW(R) contains '=', therefore, ACTION(2, =) = r_5, that is, reduce by R → L.
- Thus, we have two actions shift and reduce for (2, =) in SLR table, which means shift-reduce conflct occurs. Therefore, the grammar is not SLR(1), even if the grammar is unambiguous.

The parsing table for the above grammar is designed below:

	Action				Goto		
State	id	*	=	$	S	L	R
0	S_5	S_4			1	2	3
1				acc			
2			S_6r_5	r_5			
3				r_2			
4	S_5	S_4				8	7
5			r_4	r_4		8	9
6	S_5	S_4					
7			r_3	r_3			
8			r_5	r_5			
9				r_1			

24. Consider the following grammar G:

```
E → E + T
E → T
T → T * F
T → F
F → (E)
F → id
```

(i) List the canonical collection of sets of LR(0) items for the given grammar.

(ii) Construct SLR parsing table for the grammar.

(iii) Show the moves of the parser for the input string id * id + id.

Ans: The augmented grammar G′ for the above grammar G will be

```
E′→ E
E → E + T
E → T
T → T * F
T → F
F → (E)
F → id
```

(i) The item sets for the new grammar G′ will be determined as follows:

Item set number 0:

```
E′→ .E
+E → .E + T
+E → .T
+T → .T * F
+T → .F
+F → .(E)
+F → .id
```

In I_0, symbols just after the dot are E, T, F, (, id.

Item set number 1, I_1 (for the symbol E of I_0), we have:

```
E′→ E.
E → E. + T
```

Item set number 2, I_2 (for the symbol T of I_0), we have:

```
E → T.
T → T. * F
```

Item set number 3, I_3 (for the symbol F of I_0), we have:

```
T → F.
```

Item set number 4, I_4 (for the symbol (of I_0), we have:

```
F → (.E)
+E → .E + T
+E → .T
+T → .T * F
+T → .F
+F → .(E)
+F → .id
```

Item set number 5, I_5 (for the symbol id of I_0), we have:

```
F → id.
```

In I_1, symbol just after the dot is +.

Thus, **item set number 6**, I_6 (for the symbol '+' of I_1), we have:

```
 E → E + .T
+T → .T * F
+T → .F
+F → .(E)
+F → .id
```

In I_2, symbol after the dot is *
Thus, **item set number 7**, I_7 (for the symbol '*' of I_2), we have:

```
 T → T * .F
+F → .(E)
+F → .id
```

In I_3, there is no symbol after the dot.
In I_4, symbols after the dot are E, T, F, (, id.
Here,

T is already processed in I_2.
F is already processed in I_3.
(is already processed in I_4
id is already processed in I_5.

Thus, **item set number 8**, I_8 (for the symbol E of I_4), we have:

```
F → (E.)
```

In I_5, there is no symbol to be processed after the dot.
In I_6, symbols after the dot are T, F, (, id.
Here,

F is already processed in I_3.
(is already processed in I_4
id is already processed in I_5.

Thus, **item set number 9**, I_9 (for the symbol T of I_6), we have:

```
E → E + T.
```

In I_7, symbols after the dot are F, (, id.
Here,

(is already processed in I_4
id is already processed in I_5.

Thus, **item set number 10**, I_{10} (for the symbol F of I_7), we have:

```
T → T * F.
```

In I_8, symbol after the dot is).
Thus, **item set number 11**, I_{11} (for the symbol) of I_8), we have:

```
F → (E).
```

Now, in I_9, I_{10}, I_{11} there is no symbol left after the dot.
Therefore, the transition table will be:

Item Set	id	+	*	(	)	E	T	F
0	5			4		1	2	3
1		6						
2			7					
3								
4	5			4		8	2	3
5								
6	5			4			9	3
7	5			4				10
8					11			
9								
10								
11								

Now, the action/goto table will be designed as given below:

State	Action						Goto		
	id	+	*	(	)	$	E	T	F
0	S_5			S_4			1	2	3
1		S_6				acc			
2	r_2	r_2	S_7, r_2	r_2	r_2	r_2			
3	r_4	r_4	r_4	r_4	r_4	r_4			
4	S_5			S_4			8	2	3
5	r_6	r_6	r_6	r_6	r_6	r_6			
6	S_5			S_4				9	3
7	S_5			S_4					10
8					S_{11}				
9	r_1	r_1	r_1	r_1	r_1	r_1			
10	r_3	r_3	r_3	r_3	r_3	r_3			
11	r_5	r_5	r_5	r_5	r_5	r_5			

It is clear from the table that action entry for state 2 contains shift-reduce conflict; thus, it is not LR(0).

(ii) **For SLR parsing table:**

`FOLLOW(E) = {$}` as it is the start symbol
`Follow(E) = FIRST(+T) = {+}`
`Follow(E) = FIRST( ) ) = { ) }`
Therefore, `FOLLOW(E) = {+, ), $}`
And in state 2, reduction r_2 is valid only in the columns `{+, ), $}`

Now, `FOLLOW(T) = FOLLOW(E) = {+, ), $}`
`FOLLOW(T) = FIRST(*F) = {*}`
Therefore, `FOLLOW(T) = {+, *,), $}`
And in state 3, reduction r_4 is valid only in the columns `{+, *,), $}`

Now, `FOLLOW(F) = FOLLOW(T) = {+, *,), $}`

Therefore, in state 5 reduction r_6 is valid only in the columns {+, *,), $}, in state 9 r_1 is valid only in the columns {+,), $}, in state 10 r_3 is valid only in the columns {+, *,), $} and in state 11 r_5 is valid only in the columns {+, *,), $}.

Now, after solving shift-reduce conflict, the SLR parsing table will be:

State	Action						Goto		
	id	+	*	(	)	$	E	T	F
0	S_5			S_4			1	2	3
1		S_6				Acc			
2		r_2	S_7		r_2	r_2			
3		r_4	r_4		r_4	r_4			
4	S_5			S_4			8	2	3
5		r_6	r_6		r_6	r_6			
6	S_5			S_4				9	3
7	S_5			S_4					10
8		S_6				S_{11}			
9		r_1	S_7		r_1	r_1			
10		r_3	r_3		r_3	r_3			
11		r_5	r_5		r_5	r_5			

(iii) Moves of parser to accept the input string `id * id + id` are shown below:

Stack	Input	Action of Parser
[0]	id * id + id $	Shift '0' initially
[0 id 5]	* id + id $	Action (0, id) = S_5
[0 F 3]		Action (5, *) = r_6 Goto(0, F) = 3
[0 T 2]	(id + id $	Action (3, *) = r_4 Goto (0, T) = 2
[0 + 2 * 7]	id + id $	Action(2, *) = S_5
[0 T 2 * 7 id 5]	+ id #	Action(7, id) = S_5
[0 T 2 * 7 F 10]	+ id $	Action (5, +) = r_6 Goto(7, F) = 10
[0 T 2]	+ id $	Action (10, +) = r_3 Goto(0, T) = 2
[0 E 1]	+ id $	Action (2, +) = r_2 Goto(0, E) = 1
[0 E 1 + 6]	id $	Action (1, +1) = S_6
[0 E 1 + 6 F 3]	$	Action (5, $) = r_6 Goto(6, F) = 3
[0 E 1 + 6 T 9]	$	Action (3, $) = r_4 Goto(6, T) = 9
[0 E 1]	$	Action (9, $) = r_1 Goto(0, E) = 1
		"Acc" because Action (1, $) = "acc"

25. Construct the LR(1) items and the CLR parsing table for the following grammar:

S → CC
C → cC
C → d

Ans: The augmented grammar G′ for the above grammar will be:

```
S′→ S
S → CC
C → cC
C → d
```

Apply ITEMS(G′) procedure to construct LR(1) items.
For I_0, Closure (S′ → .S, $)

S′	→	ε	.	S	ε	,	$
A	→	α	.	B	β	,	a

Here B = S, β = ε
FIRST($) = {$} and we have the production S → CC which is of the form B → γ so add [B → .γ, b] to I, for each b in FIRST(βγ), that is, S → .CC, $ and then again computing the closure.

Closure[S → .CC, $]

S	→	ε	.	C	C	,	$
A	→	α	.	B	β	,	a

Here B = C and β = C, a = $
FIRST(βa) = FIRST(C$) = {c, d} and we have B → γ, that is, C → cC and C → d so, we add the following productions for each FIRST(βa):

```
C → .cC, c
C → .cC, d
C → .d, c
C → .d, d
```

Now, we can write LR(1) items for I_0 as:

```
S′→ .S, $
S → .CC, $
C → .cC, c/d
c → .d, c/d
```

I_1 = GOTO (I_0, s)
= Closure (S′ → S., $)
Thus, I_1 will have: S′ → S., $

I_2 = GOTO (I_0, C)
= Closure (S → C.C, $)

Thus, the productions in I_2 will be:

```
S → C.C, $
C → c.C, $   [B → .γ, a] of closure(1)
C → .d, $    [B → .γ, a] of closure(1)
```

I_3 = GOTO (I_0, c)
= Closure (C → c.C, c/d)

Thus, the productions in I_3 will be:

```
C → c.C, c/d
C → .cC, c/d   [B → .γ, a] of closure(1)
C → .d, c/d    [B → .γ, a] of closure(1)
```

I_4 = GOTO(I_0, d)
= Closure (C → d., c/d)

Thus I_4 will have: C → d., c/d

There is no transition from I_1.

I_5 = GOTO (I_2, C)
= Closure (S → CC., $)

Thus, I_5 will have:

```
S → CC., $
```

I_6 = GOTO (I_2, c)
= Closure (C → c.C, $)

Thus, the productions in I_6 will be:

```
C → c.C, $
C → .cC, $
C → .d, $
```

I_7 = GOTO (I_2, d)
= Closure (C → d., $)

Thus, I_7 will have: C → d., $

I_8 = GOTO (I_3, C)
= Closure (C → cC., c/d)

Thus, I_8 will have:

```
C → cC., c/d
```

And GOTO (I_3, C) = Closure (C → c.C, c/d)
= I_3

And GOTO (I_3, d) = Closure (C → d., c/d)
= I_4

No transition on I_4 and I_5.

So, I_9 = GOTO (I_6, C)
= Closure (C → cC., $)

Thus, I_9 will have:

```
C → cC., $
```

And, GOTO (I_6, c) = Closure (C → c.C, $)
= I_6
And, GOTO (I_6, d) = Closure (C → d., $)
= I_7
And finally, there are no transitions from states I_7, I_8, I_9.
Now, the canonical LR parser table for the above LR(1) items will be designed as follows:

State	Action			GOTO	
	c	d	$	S	C
0	S_3	S_4		1	2
1			acc		
2	S_6	S_7			5
3	S_3	S_4			8
4	r_3	r_3			
5			r_1		
6	S_6	S_7			9
7			r_3		
8	r_2	r_2			
9			r_2		

26. Discuss algorithms for computation of sets of LR(1) item. Also, show that the following grammar is LR(1) but not LALR(1).

G: S → Aa / bAc / BC / bBa
A → d
B → d

Ans: The augmented grammar G′ for the above grammar will be:

```
S′→ S
S → Aa
S → bAc
S → Bc
S → bBa
A → d
B → d
```

Item set number 0, I_0:

```
 S′→  .S, $
+S →  .Aa, $
+S →  .bAc, $
+S →  .Bc, $
+S →  .bBa, $
+A →  .d, a
+N →  .d, c
```

Symbols found in I_0 are S, A, b, B, d

Item set number 1, I_1 (for the symbol S of I_0)

```
S' → S., $
```

Item set number 2, I_2 (for the symbol A of I_0)

```
S → A. a, $
```

Item set number 3, I_3 (for the symbol b of I_0)

```
 S → b.Ac, $
 S → b.Ba, $
+A → .d, c
+B → .d, a
```

Item set number 4, I_4 (for the symbol B of I_0)

```
S → B.c, $
```

Item set number 5, I_5 (for the symbol d of I_0)

```
A → d., a
N → d., c
```

Symbol found in I_2 is a

Item set number 6, I_6 (for the symbol a of I_2)

```
S → Aa., $
```

Symbols found in I_3 are A, B, d

Item set number 7, I_7 (for the symbol A of I_3)

```
S → bA.c, $
```

Item set number 8, I_8 (for the symbol B of I_3)

```
S → bB.a, $
```

Item set number 9, I_9 (for the symbol d of I_3)

```
A → d., c
B → d., a
```

Symbol found in I_4 is c

Item set number 10, I_{10} (for the symbol c of I_4)

```
S → Bc., $
```

Symbol found in I_7 is c

Item set number 11, I_{11} (for the symbol c of I_7)

```
S → bAc., $
```

Symbol found in I_8 is a

Item set number 12, I_{12} (for the symbol a of I_8)

```
S → bBa., $
```

The action/goto table will be designed as follows:

	Action					Goto		
State	A	b	bc	d	$	S	A	B
0		S_3		S_5		1	2	4
1					acc			
2	S_6							
3				S_9			7	8
4			S_{10}					
5	r_5		r_6					
6					r_1			
7			S_{11}					
8	S_{12}							
9	r_6		r_5					
10					r_3			
11					r_2			
12					r_4			

Since the table does not have any conflict, it is LR(1).
For LALR(1) table, item set 5 and item set 9 are same.
Thus, we merge both the item sets (I_5, I_9) = item set 59, I_{59}
Now, the resultant parsing table becomes:

	Action					Goto		
State	a	b	c	d	$	S	A	B
0		S_3		S_5		1	2	4
1					acc			
2	S_6							
3				S_9			7	8
4			S_{10}					
59	r_5, r_6		r_6, r_5					
6					r_1			
7			S_{11}					
8	S_{12}							
10					r_3			
11					r_2			
12					r_4			

Since the table contains reduce-reduce conflict, it is not LALR(1).

Multiple-Choice Questions

1. The most common non-backtracking shift-reduce parsing technique is known as __________.

   (a) LL parsing (b) LR parsing
   (c) Top-down parsing (d) Bottom-up parsing

2. The simplest LR parsing technique is __________.
   (a) CLR parser
   (b) SLR parser
   (c) LALR parser
   (d) LL parser
3. `X → A.BC`, the given item indicates that __________.
   (a) a string derivable from `ABC` is expected next on the input.
   (b) a string derivable from `BC` has already been seen and now a string derivable from `A` is expected on the input.
   (c) a string derivable from `A` has already been seen on the input and now a string derivable from `BC` is expected.
   (d) the body of the production has already been seen, and now it is time to reduce it to `X`.
4. Shift-reduce and reduce-reduce conflicts occur in __________.
   (a) SLR parser
   (b) LALR parser
   (c) CLR parser
   (d) None of these
5. A parser that accommodates some extra information in the form of a terminal symbol, as a second component is known as __________.
   (a) SLR parser
   (b) LALR parser
   (c) CLR parser
   (d) LL parser
6. If $A \rightarrow \alpha\beta$ is a production and `a` is terminal symbol or right end marker `$`, then LR(1) items will be defined by the production __________.
   (a) $[A \rightarrow \alpha.\beta,\ a]$
   (b) $[A \rightarrow .\alpha\beta,\ a]$
   (c) $[A \rightarrow \alpha.\beta a]$
   (d) $[A \rightarrow a.\beta,\ \alpha]$
7. __________ parsers are specialized form of LR parsers that lie in between SLR parsers and canonical LR parsers in terms of power of parsing grammars.
   (a) LALR parser
   (b) LR(0) parser
   (c) CLR(1) parser
   (d) LR(1) parser
8. The automatic tool/tools to generate an LR parser from a given grammar is/are __________.
   (a) YACC
   (b) LEX
   (c) BISON
   (d) Both (a) and (b)

Answers

1. (b) 2. (b) 3. (c) 4. (a) 5. (c) 6. (a) 7. (a) 8. (d)

6

Syntax-directed Translations

1. What is syntax-directed translation? Explain.

Ans: Syntax-directed translation (SDT) is an extension of context-free grammar (CFG) which acts as a notational framework for the generation of intermediate code. It embeds program fragments (called **semantic actions**) within the production bodies. For example, as in

$$A \rightarrow A_1 * B \ \{\texttt{print '*'}\}$$

In SDT, by convention, we use curly braces to enclose semantic actions. If the grammar symbols contain curly braces, then we enclose them in single quotes as `'{'` and `'}'`. The order of execution of semantic actions is determined by the position of a semantic action in a production body. Here, in the above example, the action occurs at the end, that is, after all the grammar symbols. However, semantic actions can occur at any position in the production body.

In syntax-directed translation, we pass the token stream as input, build the parse tree for the input token stream, and then evaluate the semantic rules at the nodes of parse tree by traversing the parse tree (as shown in Figure 6.1). The evaluation of semantic rules by syntax-directed translation may generate an intermediate code, save information in the symbol table, perform semantic checking, issue error messages, or perform some other activities.

Input string ⟶ Parse tree ⟶ Dependency Graph ⟶ Evaluation order for semantic rules

Figure 6.1 Conceptual View of Syntax Directed Translation

2. Explain syntax-directed definition.

Ans: Syntax-directed definition (SDD) associates attributes with the grammar symbols and semantic rules with each grammar production. In other words, SDD is a context-free grammar with the information to control the semantic analysis and translation. These grammars are also called **attribute grammars**. A grammar is augmented by associating attributes with each grammar symbol that describes its properties, and the rules are associated with productions. If `A` is a symbol and `i` is one of its attributes, then we can write `A.i` to denote the value of `i` at a particular parse tree node. An attribute has a name and an associated value which can be a string, a number, a type, a memory location, or an assigned register. The attribute value may depend on its child nodes or siblings or its parent node information.

The syntax-directed definition is partitioned into two subsets, namely, *synthesized* and *inherited* attributes. Semantic rules are used to set up dependencies between attributes that will be represented by a graph. An evaluation order for the semantic rules can be derived from the dependency graph. The values of the attributes at the nodes in parse tree are defined by the evaluation of the semantic rules.

3. Compare syntax-directed translation and syntax-directed definition.

Ans: Translations are carried out during parsing and the order in which the semantic rules are evaluated by the parser must be explicitly specified. Hence, instead of using the syntax-directed definitions, we use syntax-directed translation schemes to specify the translations. Syntax-directed definitions are more abstract specifications for translations; therefore, they hide many implementation details, freeing the user from having to explicitly specify the order in which translation takes place. Whereas, the syntax-directed translation schemes indicate the order in which semantic rules are evaluated, allowing some implementation details to be specified. Syntax-directed definitions are more readable and therefore, are useful for specification, whereas translation schemes are very much efficient and, therefore, are useful for implementation.

4. What are synthesized and inherited attributes? How semantic rules are attached to the productions?

Ans: Synthesized attributes: A synthesized attribute for a non-terminal `A` at a parse tree node `I` is defined by a semantic rule associated with the production at `I` and the production must have `A` as its head. A synthesized attribute at node `I` has its value defined only in terms of attribute values of its child nodes and `I` itself.

For example, consider the grammar for a desk calculator.

```
E → M
M → M′ + P
M → P
P → P′ * R
P → R
R → (M)
R → digit
```

The given grammar describes arithmetic expressions with operators + and *. In SDD, each of the non-terminals has a single synthesized attribute, called `val`, and `digit` has a synthesized attribute `lexval` (the lexical value for the digit which is an integer value returned by the lexical analyzer).

So, semantic rules for this grammar can be written as given in Figure 6.2.

In production rule 1, `E → M`, sets `E.val` to `M.val`. The production rule 2, `M → M′ + P`, computes the `val` attribute for the head `M` as the sum of the values at its child nodes `M′` and `P`. Production 3, 5, and 6 are same as that of production rule 2. Production rule 7 gives `R.val`, the numerical value of the token digit which is returned by the lexical analyzer.

Since the entire attribute values of the symbols as head are defined in terms of attributes at their child nodes, it means all attributes involved are synthesized attributes and the corresponding SDD is known as **S-attributed definition**.

PRODUCTION	SEMANTIC RULES
1. E → M	E.val = M.val
2. M → M′ + P	M.val = M′.val + P.val
3. M → P	M.val = P.val
4. P → P′*R	P.val = P′.val * R.val
5. P → R	P.val = R.val
6. R → (M)	R.val = M.val
7. R → digit	R.val = digit.lexval

Figure 6.2 Syntax Directed Definition of a Simple Desk Calculator

Inherited attributes: An inherited attribute for a non-terminal A at a parse tree node I is defined by a semantic rule associated with the production at the parent of I, and the production must have A as a symbol in its body. The value of an inherited attribute at node I can only be defined in terms of attribute values of I's parents, I's siblings, and I itself. Inherited attributes are convenient for expressing the dependence of a programming language construct on the context on which it appears. For example, an inherited attribute can be used to keep track of whether an identifier appears on the left side or right side of an assignment operator in order to determine whether the address or the value of the identifier is required.

For example, consider the following grammar:

```
E → AB
A → int
B → B′,id
B → id
```

The syntax-directed definitions that use inherited attributes can be written as:

PRODUCTION	SEMANTIC RULES
1. E → AB	B.inh = A.type
2. A → int	A.type = integer
3. B → B′,id	B′.inh = B.inh enter (id.print, B.inh)
4. B → id	enter (id.print, B.inh)

Figure 6.3 Syntax Directed Definition for Inherited Attributes

The non-terminal symbol A in the productions has the synthesized attribute type whose value can be obtained by the keyword in the declaration. The semantic rule B.inh = A.type sets inherited attributes B.inh to the type in the declaration.

The parse tree with the attributes values at the parse tree nodes, for an input string int id_1, id_2, id_3 is shown in Figure 6.4. The type of identifiers id_1, id_2, and id_3 is determined by the value of B.inh at the three B nodes. These values are obtained by computing the value of the attribute A.type at the left child of the root and then evaluating B.inh in top-down at the three B nodes in the right subtree of the root. We also call the procedure enter at each B node to insert into the symbol table, where the identifier at the right child of this node is of type int.

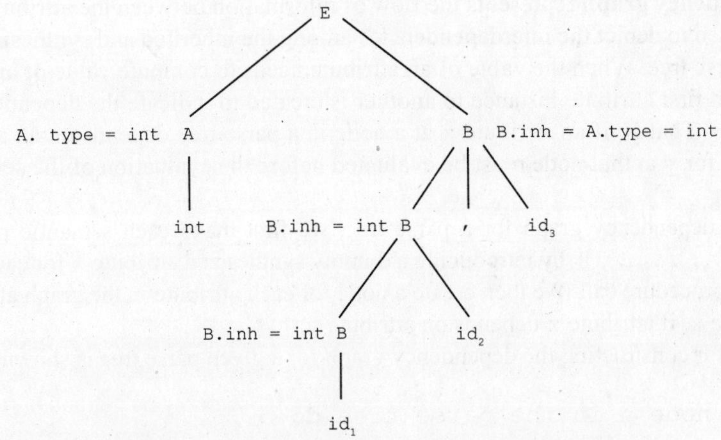

Figure 6.4 Parse Tree for String Int id_1, id_2, id_3 with Inherited Attributes

5. Define annotated parse tree with example.

Ans: An **annotated parse tree** is a parse tree that displays the values of the attributes at each node. It is used to visualize the SDD specified translations. To construct an annotated parse tree, first the SDD rules are applied to construct a parse tree and then the same SDD rules are applied to evaluate the attributes at each node of the parse tree. For example, if all the attributes are synthesized then we must evaluate the attribute values of all the children of a node before evaluating the attribute value of the node itself.

For example, an annotated parse tree for an expression `3 + 5 * 2` by considering the productions and semantic rules of Figure 6.2, is shown in Figure 6.5.

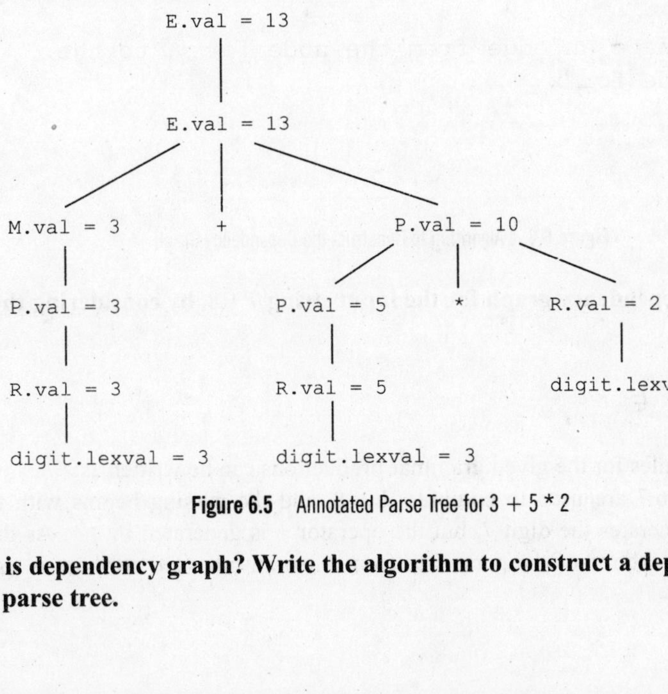

Figure 6.5 Annotated Parse Tree for 3 + 5 * 2

6. What is dependency graph? Write the algorithm to construct a dependency graph for a given parse tree.

Ans: A **dependency graph** represents the flow of information between the attribute instances in a parse tree. It is used to depict the interdependencies among the inherited and synthesized attributes at the nodes in a parse tree. When the value of an attribute needs to compute value of another attribute, then an edge from first attribute instance to another is created to indicate the dependency among the attribute instances. That is, if an attribute x at a node in a parse tree depends on an attribute y, then the semantic rule for y at that node must be evaluated before the evaluation of the semantic rule that defines x.

To construct a dependency graph for a parse tree, we first make each semantic rule in the form `x:= f(y1,y2,y3,...,yk)` by introducing a dummy synthesized attribute x for each semantic rule that consists of a procedure call. We then create a node for each attribute in the graph and an edge from node y to the node x, if attribute x depends on attribute y.

The algorithm for constructing the dependency graph for a given parse tree is shown in Figure 6.6.

```
For each node n in the parse tree do
Begin
    For each attribute b of the grammar symbol at n do
    Begin
        construct a node in the dependency graph for b
    End
End
For each node n in the parse tree do
Begin
    For each semantic rule x:= f(y1, y2,...yk) associated with
    the production used at n do
    Begin
        For i:= 1 to k do
        Begin
            create an edge from the node for yi to the
            node for x
        End
    End
End
```

Figure 6.6 Algorithm to Construct the Dependency Graph

7. Construct a dependency graph for the input string 7 + 8, by considering the following grammar:

A → BA′
A′ → + BA₁′| ∈
B → digit

Ans: The semantic rules for the given grammar productions can be written as:

The SDD in Figure 6.7 are used to compute 7 + 8, and the parsing begins with the production A → BA′. Here, B generates the digit 7, but the operator + is generated by A′. As the left operand 7 appears in a different subtree of the parse tree from +, so an inherited attribute is used to pass the operand to the operator.

PRODUCTION	SEMANTIC RULES
1. A → BA′	A′.inh = B.val A.val = A′.syn
2. A′→ + BA$_1$′	A$_1$′.inh = A′.inh + B.val A′.syn = A$_1$′.syn
3. A′ → ∈	A′.syn = A′.inh
4. B → digit	B.val = digit.lexval

Figure 6.7 Syntax Directed Definition for the Given Grammar

A synthesized attribute `val` is used for each of the non-terminals A and B and a synthesized attribute `lexval` is used for the terminal `digit`. The non-terminal A′ has two attributes, an inherited attribute `inh` and a synthesized attribute `syn`.

In the given string 7 + 8, the operator + inherits 7 as shown in Figure 6.8.

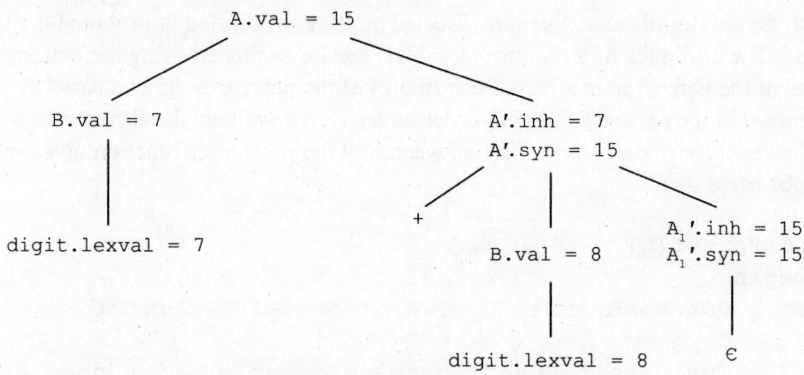

Figure 6.8 Annotated Parse Tree for 7 + 8

We use this parse tree to construct a dependency graph and the corresponding dependency graph is shown in Figure 6.9. The nodes in the dependency graph are numbered from 1 to 9, and they correspond to the attributes in the annotated parse tree.

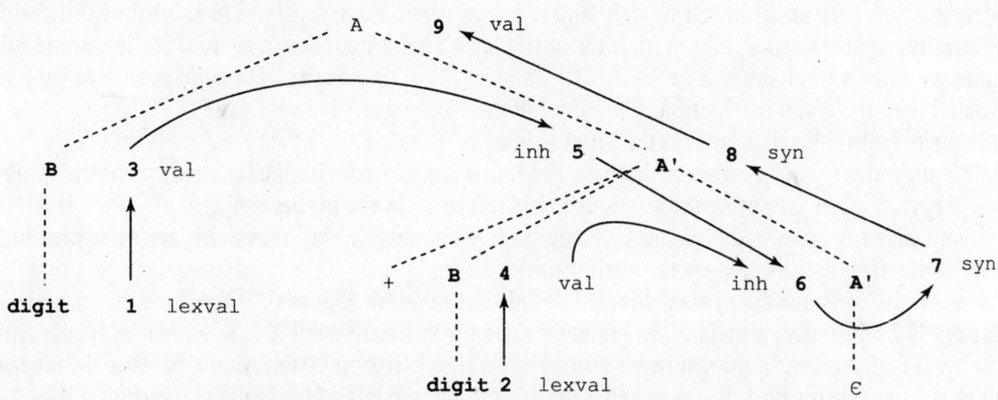

Figure 6.9 Dependency Graph for the Annotated Parse Tree of Figure 6.8

The two leaves `digit` are associated with attribute `lexval` and are represented by nodes 1 and 2. The two nodes labeled `B` are associated with the attribute `val` and are represented by the nodes 3 and 4. The edges from node 1 to node 3 and from node 2 to node 4 use the semantic rule that defines `B.val` in terms of `digit.lexval`.

Each occurrence of non-terminal `A′` is associated with the inherited attribute `A′.inh`, and are represented by nodes 5 and 6. The edge from 3 to 5 is due to the rule `A′.syn` = `A′.inh`. The edge from node 5 to node 6 is for `A′.inh` and from node 4 to node 6 is for `B.val`, because these values are added to calculate the attribute `inh` at node 6.

The synthesized attribute `syn` associated with the occurrences of `A′` is represented by nodes 7 and 8. The edge from node 6 to node 7 is due to the semantic rule `A′.syn` = `A′.inh` associated with the production 3. The edge from node 7 to node 8 is due to the semantic rule associated with production 2. Node 9 represents the attribute `A.val`. The edge from node 8 to node 9 is due to the semantic rule, `A.val` = `A′.syn`, associated with production 1.

8. What are S-attributed definitions and L-attributed definitions?

Ans: S-Attributed definitions: A syntax-directed translation is called S-attributed if all its attributes are synthesized. The attributes of an S-attributed SDD can be evaluated using the bottom-up approach of the traversal of the parse tree in which the attributes of the parse tree are evaluated by performing a post-order traversal of the parse tree. In post-order traversal, we evaluate the attributes at a node `N` when the traversal leaves `N` for the last time. That is, we, apply the post-order function given in Figure 6.10 to the root of the parse tree.

```
postorder(N)
Begin
        For each child C of N, from left to right
            postorder(C);
        Evaluate the attributes associated with node N
End
```

Figure 6.10 Algorithm for Computing Postorder Function

L-attributed definition: An `L`-attributed definition is another class of SDD in which the dependency graph edges can only go from left to right and not vice-versa. Each attribute in L-attributed definition must be either synthesized or inherited. If the attributes are inherited, then they must follow some rules. Assume we have a production $A \to Y_1Y_2 \ldots Y_n$, and $Y_i.a$ is an inherited attribute evaluated by a rule associated with the given production. Then the rule may only use:

- Inherited attributes that are related with the head `A`.
- The attributes (either synthesized or inherited) that are related with the occurrences of the symbols $Y_1, Y_2, \ldots, Y_{i-1}$ (that is, the symbols to the left of Y_i in the production).
- Synthesized or inherited attributes related with Y_i in such a way that there are no cycles in the dependency graph formed by the attribute of Y_i.

For example, consider the syntax-directed definitions given in Figure 6.7. To prove that the SDD in Figure 6.7 is L-attributed, consider the semantic rules for inherited attributes as shown in Figure 6.11.

The syntax-directed definition above is an example of the L-attributed definition, because the inherited attribute `A′.inh` using only `B.val`, and `B` is appearing to the left of `A′` in the production `A` → `BA′`. Similarly, the inherited attribute A_1`′.inh` in the second rule is defined using the inherited attribute

PRODUCTION	SEMANTIC RULES
A → BA′	A′.inh = B.Val
A → +BA$_1$′	A$_1$′.inh = A′.inh + B.val

Figure 6.11 Syntax Directed Definition of Inherited Attributes.

A′.inh related with the head, and B.val, where B appears on the left of A$_1$′ in the production A′ → + BA$_1$′. In both cases, the rules for L-attributed definitions are followed, and the remaining attributes are synthesized (as shown in Figure 6.7). Therefore, this SDD is L-attributed.

9. Discuss the applications of syntax-directed translation.

Ans: Syntax-directed translations are applied in the following techniques:

- **Construction of syntax tree:** Syntax tree is used as an intermediate representation in some compilers and, hence, a common form of SDD converts its input string into the syntax tree. To construct the syntax tree for expressions, we can use either an S-attributed or an L-attributed definition. The S-attributed definitions are suitable to use in bottom-up parsing, whereas the L-attributed definitions are suitable to use in top-down parsing.
- **Type checking**: Type checking is used to catch errors in the program by checking type of each variable, constant, functions, and expressions. Type checking eliminates the need for dynamic checking for type errors.
- **Intermediate code generation**: Intermediate codes are machine-dependent codes and are close to the machine instructions. Syntax-directed translation, postfix notation, and syntax tree can be used as an intermediate code.

10. What is a syntax tree? Explain the procedure for constructing a syntax tree with the help of an example.

Ans: A **syntax tree** or an **abstract syntax tree (AST)** is a tree representation showing the syntactic structure of the source program. It is a compressed form of a parse tree representing the hierarchical construction of a program, where the nodes represent operators and the children of any node represent the operands that are to be operated by that operator. For example, the syntax tree for the expression `p * (q + r)/s` is shown as in Figure 6.12.

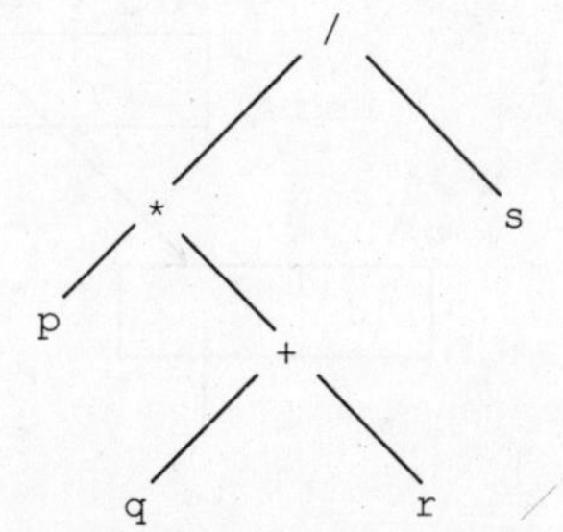

Figure 6.12 A Simple (Abstract) Syntax Tree

The construction of syntax tree for an expression can be considered as the translation of the expression into post-fix form. The subtrees are constructed for the subexpressions by creating a node for each operator and operand. The children of an operator node are the roots of the nodes representing the subexpressions constituting the operands of that operator.

The nodes of a syntax tree are implemented as objects having several fields. Each node is labeled by the `op` field, which is often called the label of the node. When used for translation, the nodes in a syntax tree may have additional fields to hold the values of attributes attached to the node, which are as follows:

- For a leaf node, an additional field is required to hold the lexical value of the leaf. A constructor function `Make-leaf(num, val)` or `Make-leaf(id, entry)` is used to create a leaf object.
- If the node is an interior node, a constructor function `Make-node(op,left,right)` is used to create an object with first field `op` and two additional fields for its left and right children.

For example, consider the expression `x - 7 + z`. In this, we need the following functions to create the nodes of syntax trees for expressions with binary operators.

- `Make-node (op, left, right)` creates an operator node with label `op` and two fields containing pointers to `left` and `right` children.
- `Make-leaf (id, entry)` creates an identifier node with label `id` and a field containing `entry`, a pointer to the symbol table entry for the identifier.
- `Make-leaf (num, val)` creates a number node with label `num` and a field containing `val`, the value of the number.

Consider the S-attributed definition shown in Figure 6.13 that constructs syntax tree for the expressions involving only binary operators + and –. All the non-terminals have only one synthesized attribute `node` that represents a node of the syntax tree.

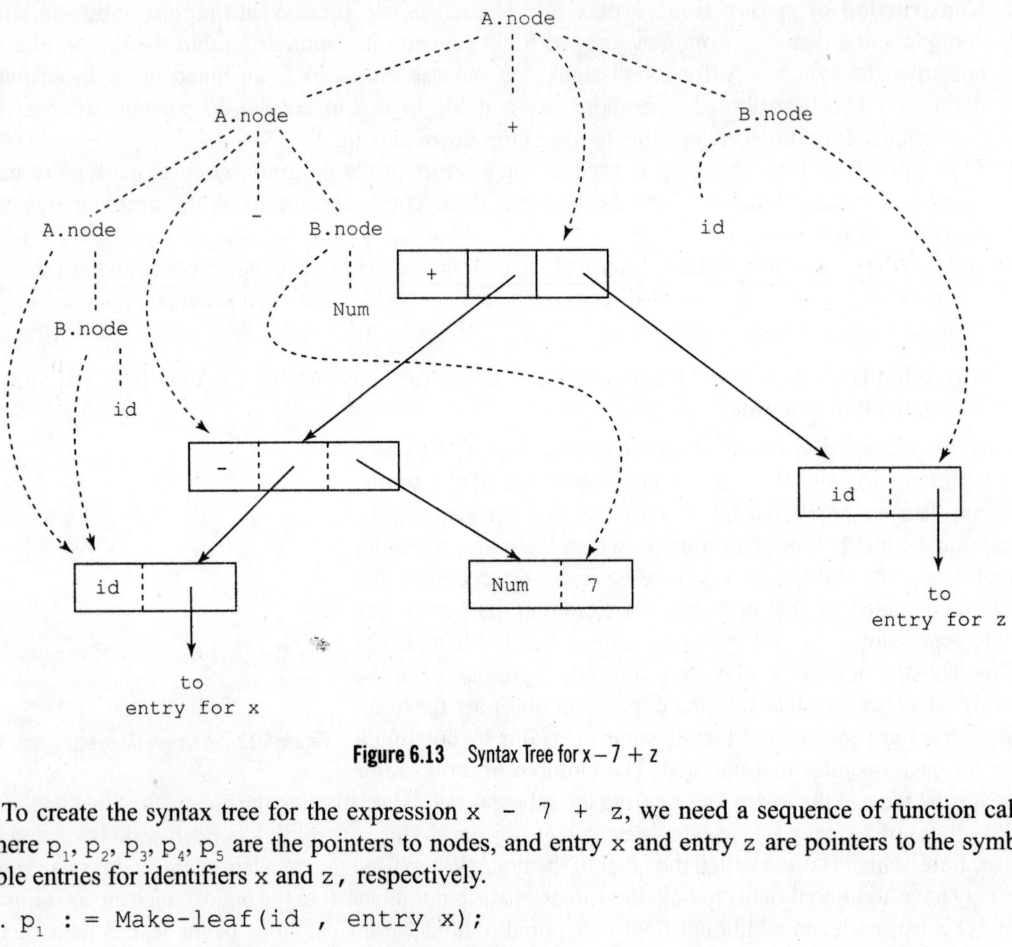

Figure 6.13 Syntax Tree for x – 7 + z

To create the syntax tree for the expression `x - 7 + z`, we need a sequence of function calls, where p_1, p_2, p_3, p_4, p_5 are the pointers to nodes, and entry `x` and entry `z` are pointers to the symbol table entries for identifiers `x` and `z`, respectively.

1. p_1 : = Make-leaf(id , entry x);
2. p_2 : = Make-leaf(num , 7);
3. p_3 : = Make-leaf(' - ', p_1 , p_2);
4. p_4 : = Make-leaf(id , entry z);
5. p_5 : = Make-leaf('+' , p_3 , p_4);

The tree is constructed using bottom-up approach. The function calls `Make-node(id, entry x)` and `Make-node(num,7)` construct the leaves for `x` and 7, the pointers to these nodes are saved using p_1 and p_2.

The function call `Make-node('-', p1, p2)` constructs an interior node with the leaves for `x` and 7 as children and we follow the same procedure for pointer p_4 and p_5, which finally results in p_5 pointing to the root of the constructed syntax tree.

The edges of the syntax tree are shown as solid lines. The underlying parse tree is shown with dotted lines and the dashed lines represent the values of `A.node` and `B.node`, each line points to appropriate node of the syntax tree.

11. What do you understand by syntax-directed translation schemes?

Ans: Syntax-directed translation scheme is an extension of context-free grammar with program fragments (known as **semantic actions**) embedded within production bodies. The semantic actions are generally enclosed within the curly braces, and if the braces are needed as grammar symbols they are put in single quotes. The semantic actions can be placed at any position within a production body. Syntax-directed translation schemes can be considered as a complementary notation to SDD.

Syntax-directed translations can be implemented by first constructing a parse tree and then performing the actions in a left-to-right depth-first order, that is, during a preorder traversal. A syntax-directed translation scheme having both synthesized and inherited attributes needs to be careful while translating them and must follow the given rules:

- An inherited attribute for a symbol on the right side of a production must be computed in an action before that symbol.
- A synthesized attribute of a symbol to the right of the action would not be referred by an action.
- A synthesized attribute for the non-terminal on the left can only be computed after all attributes if references have been computed. The action for computing such attributes can usually be placed at the end of the right side of the production.

Multiple-Choice Questions

1. Which of the following is not true for SDT?
   (a) It is an extension of CFG.
   (b) Parsing process is used to do the translation.
   (c) It does not permit the subroutines to be attached to the production of a CFG.
   (d) It generates the intermediate code.

2. A parse tree with attribute __________ at each node is known as an annotated parse tree.
   (a) Name (b) Value
   (c) Label (d) None of these

3. Which of the following is true for a dependency graph?
   (a) The dependency graph helps to determine how the attribute values are computed.
   (b) It depicts the flow of information among the attribute instances in a parse tree.
   (c) Both (a) and (b)
   (d) None of these

4. An SDD is S-attributed if every attribute is __________.
   (a) Inherited (b) Synthesized
   (c) Dependent (d) None of these

5. In L-attributed definitions, the dependency graph edges can go from _________ to _________.
   (a) Left to right
   (b) Right to left
   (c) Top to bottom
   (d) Bottom to top
6. Which of the following is not true for an abstract syntax tree?
   (a) It is a compressed form of a parse tree.
   (b) It represents the syntactic structure of the source program.
   (c) The nodes of the tree represent the operands.
   (d) None of these
7. Which of the following is not true for syntax-directed translation schemes?
   (a) It is a CFG with program fragments embedded within production bodies.
   (b) The semantic actions appear at a fixed position within a production body.
   (c) They can be considered as a complementary notation to syntax-directed definitions.
   (d) None of these

Answers

1. (c) 2. (b) 3. (c) 4. (b) 5. (a) 6. (c) 7. (b)

7

Intermediate Code Generation

1. What is intermediate code?

Ans: During the translation of a source program into the object code for a target machine, a compiler may generate a middle-level language code, which is known as **intermediate code** or **intermediate text**. The complexity of this code lies between the source language code and the object code. The intermediate code can be represented in the form of postfix notation, syntax tree, directed acyclic graph (DAG), three-address code, quadruples, and triples.

2. Write down the benefits of using an intermediate code generation over direct code generation?

Ans: The benefits of using an intermediate code over direct code generation are as follows:

- Intermediate code is machine independent, which makes it easy to retarget the compiler to generate code for newer and different processors.
- Intermediate code is nearer to the target machine as compared to the source language so it is easier to generate the object code.
- The intermediate code allows the machine-independent optimization of the code. Several specialized techniques are used to optimize the intermediate code by the front end of the compiler.
- Syntax-directed translation implements the intermediate code generation; thus, by augmenting the parser, it can be folded into the parsing.

3. What are the two representations to express intermediate languages?

Ans: The two representations of intermediate languages are categorized as follows:

- **High-level intermediate representation:** This representation is closer to the source program. Thus, it represents the high-level structure of a program, that is, it depicts the natural hierarchical structure of the source program. The examples of this representation are directed acyclic graphs (DAG) and syntax trees. This representation is suitable for static type checking task. The critical features of high-level representation are given as follows:

- It retains the program structure as it is nearer to the source program.
- It can be constructed easily from the source program.
- It is not possible to break the source program to extract the levels of code sharing due to which the code optimization in this representation becomes a bit complex.

❑ **Low-level intermediate representation:** This representation is closer to the target machine where it represents the low-level structure of a program. It is appropriate for machine-dependent tasks like register allocation and instruction selection. A typical example of this representation is three-address code. The critical features of low-level representation are given as follows:

- It is near to the target machine.
- It makes easier to generate the object code.
- High effort is required by the source program to generate the low-level representation.

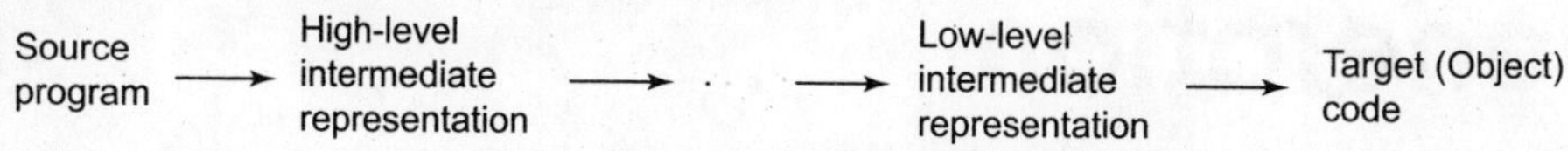

Figure 7.1 A Sequence of Intermediate Representation

4. What is postfix notation? Explain with example.

Ans: Generally, we use infix notation to represent an arithmetic expression such as multiplication of two operands `a` and `b`. In infix notation, operator is always placed between the two operands, as `a * b`. But in **postfix notation** (also known as **reverse polish** or **suffix notation**), the operator is shifted to the right end, as `ab*`. In postfix notation, parentheses are not required because the position and the number of arguments of the operator allow only a single way of decoding the postfix expression. The postfix notation can be applied to `k`-ary operators for any `k > 1`. If β is a `k`-ary operator and $a_1, a_2, \ldots, a_k$ are any postfix expressions, then after applying β to the expressions, the expression in postfix notation is represented as $a_1\ a_2 \ldots a_k\beta$.

For example, consider the following infix expressions and their corresponding postfix notations:

❑ `(l + m) * n` is an infix expression, the postfix notation will be `l m + n *`.

❑ `p * (q + r)` is an infix expression, the postfix expression will be `p q r + *`.

❑ `(p - q) * (r + s) + (p - q)` is an infix expression, the postfix expression will be `p q - r s + * p q - +`.

5. Explain the process of evaluation of postfix expressions.

Ans: The postfix notations can easily be evaluated by using a stack, and generally the evaluation process scans the postfix code left to right.

1. If the scan symbol is an operand, then it is pushed onto the stack, and the scanning is continued.
2. If the scan symbol is a binary operator, then the two topmost operands are popped from the stack. The operator is applied to these operands, and the result is pushed back to the stack.
3. If the scan symbol is an unary operator, it is applied to the top of the stack and the result is pushed back onto the stack.

Note: The result of an unary operator can be shown within parenthesis. For example, (−X).

6. Convert the following expression to the postfix notation and evaluate it.

`P + (-Q + R * S)`

Ans: The postfix notation for the given expression is:

`PQ - RS * ++`

The step-by-step evaluation of this postfix expression is shown in Figure 7.2.

S. no.	String and scan symbol	Previous stack content	Rule in use	New stack content
1.	PQ - RS * ++			
2.	P		Rule 1	P
3.	Q	P	Rule 1	PQ
4.	-	PQ	Rule 3	P(-Q)
5.	R	P(-Q)	Rule 1	P(-Q)P
6.	S	P(-Q)R	Rule 1	P(-Q)RS
7.	*	P(-Q)RS	Rule 2	P(-Q) (R * S)
8.	+	P(-Q) (R * S)	Rule 2	P(-Q + R * S)
9.	+	P(-Q + R * S)	Rule 2	P +(-Q + R * S)

Figure 7.2 Evaluation of Postfix Expression PQ - RS * ++

The desired result is P +(-Q + R * S).

7. What is a three-address code? What are its types? How it is implemented?

Ans: A string of the form X: = Y OP Z, in which op is a binary operator, Y and Z are the addresses of the operands, and X is the address of the result of the operation, is known as **three-address statement.** The operator op can be a fixed or floating-point arithmetic operator, or a logical operator. X, Y, and Z can be considered either as constants or as predefined names by the programmer or temporary names generated by the compiler. This statement is named as the "three-address statement" because of the usage of three addresses, one for the result and two for the operands. The sequence of such three-address statements is known as **three-address code**. The complicated arithmetic expressions are not allowed in three-address code because only a single operation is allowed per statement. For example, consider an expression A + B * C, this expression contains more than one operator so the representation of this expression in a single three-address statement is not possible. Hence, the three-address code of the given expression is as follows:

T_1: = B * C
T_2: = A + T_1

where, T_1 and T_2 are the temporary names generated by the compiler.

Types of Three-Address Statements:

There are some cases where a statement consists of less than three addresses and is still known as three-address statement. Hence, the different forms of three-address statements are given as follows:

- **Assignment statements:** These statements can be represented in the following forms:
  - X: = Y op Z, where op is any logical/arithmetic binary operator.
  - X: = op Y, where op is an unary operator such as logical negation, conversion operators, and shift operators.
  - X: = Y, where the value of Y is assigned to operand X.
- **Indexed assignment statements:** These statements can be represented in the following forms:
  - X: = Y[I]
  - X[I]: = Y, where X, Y and I refer to the data objects and are represented by pointers to the symbol table.

- **Address and pointer assignment statements:** These statements can be represented in the following forms:
  - `X: = addr Y` defines that `X` is assigned the address of `Y`.
  - `X: = *Y` defines that `X` is assigned the content of location pointed to by `Y`.
  - `*X: = Y` sets the r-value of the object pointed to by `X` to the r-value of `Y`.
- **Jump statements:** Jump statements are of two types—conditional and unconditional that works with relational operators and are represented in the following forms:
  - The unconditional jump is represented as `goto L`, where `L` being a label. This instruction means that the Lth three-address statement is the next to be executed.
  - The conditional jumps such as if `X relop Y goto L`, where `relop` signifies the relational operator (≤, =, >, etc.) applied between `X` and `Y`. This instruction implies that if the result of the expression `X` relop `Y` is true then the statement labeled `L` is executed. Otherwise, the statement immediately following the `if X relop Y goto L` is executed.
- **Procedure call/return statements:** These statements can be defined in the following forms:
  - `param` **`X`** and **`call P, n,`** where they are represented and typically used in the three-address statement as follows:

```
param X1
param X2
   .
   .
   .
param Xn
call P, n
```

Here, the sequence of three-address statements is generated as a part of call of the procedure `P(`X_1, X_2`, . . . ,` X_n`)`, and `n` in `call P, n` is defined as an integer specifying the total number of actual parameters in the call.

  - `Y = call p, n` represents the function call.
  - `return Y,` represents the return statement, where `Y` is a returned value.

Implementation of Three-Address Statements:

The three-address statement is an abstract form of intermediate code. Hence, the actual implementation of the three-address statements can be done in the following ways:

- Quadruples
- Triples
- Indirect triples

8. Explain quadruples with the help of a suitable example.

Ans: Quadruple is defined as a record structure used to represent a three-address statement. It consists of four fields. The first field contains the operator, the second and third fields contain the operand 1 and operand 2, respectively, and the last field contains the result of that three-address statement. For better understanding of quadruple representation of any statement, consider a statement, `S = -z/a * (x + y),` where `-z` stands for unary minus `z`.

To represent this statement into quadruple representation, we first construct the three-address code as follows:

```
t1: = x + y
t2: = a * t1
t3: = - z
t4: = t3/t2
S: = t4
```

The quadruple representation of this three-address code is shown in Figure 7.3.

	Operator	Operand 1	Operand 2	Result
0	+	x	Y	t_1
1	*	a	t_1	t_2
2	–	Z		t_3
3	/	t_3	t_2	t_4
4	: =	t_4		S

Figure 7.3 Quadruple Representation for S = –z/a * (x + y)

9. Define triples and indirect triples. Give suitable examples for each.

Ans: Triples: A triple is also defined as a record structure that is used to represent a three-address statement. In triples, for representing any three-address statement three fields are used, namely, operator, operand 1 and operand 2, where operand 1 and operand 2 are pointers to either symbol table or they are pointers to the records (for temporary variables) within the triple representation itself. In this representation, the result field is removed to eliminate the use of temporary names referring to symbol table entries. Instead, we refer the results by their positions. The pointers to the triple structure are represented by parenthesized numbers, whereas the symbol-table pointers are represented by the names themselves. The triples representation of the expression (in Question 7) is shown in Figure 7.4.

	Operator	Operand 1	Operand 2
0	+	x	Y
1	*	a	(0)
2	–	z	
3	/	(2)	(1)
4	:=	S	(3)

Figure 7.4 Triple Representation for S = –z/a * (x + y)

In triple representation, the ternary operations `X[I] : = Y` and `X : = Y[I]` are represented by using two entries in the triple structure as shown in Figure 7.5(a) and (b) respectively. For the operation `X[I] : = Y`, the names `X` and `I` are put in one triple, and `Y` is put in another triple. Similarly, for the operation `X : = Y[I]`, we can write two instructions, `t: = Y[I]`, and `X: = t`. Note that instead of referring the temporary `t` by its name, we refer it by its position in the triple.

Indirect triples: An indirect triple representation consists of an additional array that contains the pointers to the triples in the desired order. Let us define an array `A` that contains pointers to triples in desired order. Indirect triple representation for the statement `S` given in the previous question is shown in Figure 7.6.

	Operator	Operand 1	Operand 2
0	[]=	X	I
1		Y	

(a) Triple Representation of X[I] : = Y

	Operator	Operand 1	Operand 2
0	=[]	Y	I
1	:=	(0)	X

(b) Triple Representation of X : = Y[I]

Figure 7.5 More Triple Representations

	A
101	(0)
102	(1)
103	(2)
104	(3)
105	(4)

	Operator	Operand 1	Operand 2
0	+	x	y
1	*	a	(0)
2	-	z	
3	/	(2)	(1)
4	:=	S	(3)

Figure 7.6 Indirect Triples Representation of $S = -z/a * (x + y)$

The main advantage of indirect triple representation is that an optimizing compiler can move an instruction by simply reordering the array A, without affecting the triples themselves.

10. Explain Boolean expressions. What are the different methods available to translate Boolean expression?

Ans: Boolean operators, like `AND(&&)`, `OR(||)` and `NOT(!)`, play an important role in constructing the Boolean expressions. These operators are applied to either relational expressions or Boolean variables. In programming languages, Boolean expressions serve two main functions, given as follows:

- Boolean expressions can be used as conditional expressions in the statements that alter the flow of control, such as in `while-` or `if-then-else` statements.
- Boolean expressions are also used to compute the logical values.

Boolean expressions can be generated by using the following grammar:

```
S → S or S|S and S|not S|(S)|id|id relop id|true|false
```

where the attribute `relop` is used to indicate any of $<, \leq, =, \neq, >, \geq$. Generally, we consider that the operators **`or`** and **`and`** are left-associative, and that the operator **`not`** has the highest precedence, then **`and`**, and then **`or`**.

Methods of Translating a Boolean Expression into Three-Address Code:

There are two methods available to translate a Boolean expression into three-address code, as given below:

- **Numerical representation:** The first method of translating Boolean expression into three-address code comprises encoding true and false numerically and then evaluating the Boolean expression similar to an arithmetic expression. True is often denoted by 1 and false by 0. Some other encodings are also possible where any non-zero or non-negative quantity indicates true and any negative or zero number indicates false. Expressions will be calculated from left to right like arithmetic expressions. For example, consider a Boolean expression X and Y or Z, the translation of this expression into three-address code is as follows:

```
t1: = X and Y
t2: = t1 or Z
```

Now, consider a relational expression if X > Y then 1 else 0, the three-address code translation for this expression is as follows:

```
1. if X > Y goto (4)
2. t1: = 0
3. goto (5)
4. t1: = 1
5. Next
```

Here, t_1 is a temporary variable that can have the value 1 or 0 depending on whether the condition is evaluated to true or false. The label Next represents the statement immediately following the else part.

- **Control-flow representation:** In the second method, the Boolean expression is translated into three-address code based on the flow of control. In this method, the value of a Boolean expression is represented by a position reached in a program. In case of evaluating the Boolean expressions by their positions in program, we can avoid calculating the entire expression. For example, if a Boolean expression is X and Y, and if X is false, then we can conclude that the entire expression is false without having to evaluate Y. This method is useful in implementing the Boolean expressions in control-flow statements such as if-then-else and while-do statements. For example, we consider the Boolean expressions in context of conditional statements such as
  - If X then S_1 else S_2
  - while X do S

  In the first statement, if X is true, the control jumps to the first statement of the code for S_1, and if X is false, the control jumps to the first statement of S_2 as shown in Figure 7.7(a). In case of second statement, when X is false, the control jumps to the statement immediately following the while statement, and if X is true, the control jumps to the first statement of the code for S as shown in Figure 7.7(b).

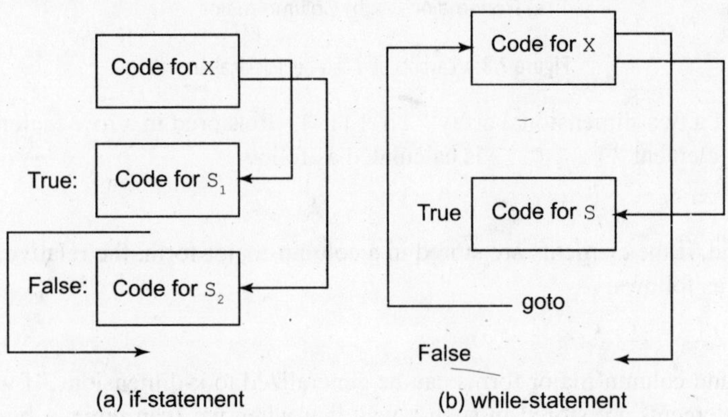

Figure 7.7 Control-flow Translation of Boolean Expressions

11. What is postfix translation?

Ans: A translation scheme is said to be postfix translation if, for each production `S` → α, the transition rule for `S.CODE` consists of the concatenation of the `CODE` translations of the non-terminals in α, in the same order as the non-terminals appear in α, followed by a tail of output. It is easy to use the postfix translation of `CODE` as it reduce space requirements, otherwise we have to follow a long scheme to construct the intermediate language form of any program like generating a parse tree followed by a walk of the tree.

12. Explain the array references in arithmetic expressions.

Or

Discuss the addressing of array elements.

Ans: An **array** is a collection of items having similar data type, which are stored in a block of consecutive memory locations. In case of languages like C and Java, array consists of `n` elements, numbered `0, 1, . . ., n - 1`. For a single-dimensional array, the address of ith element of the array is calculated as `base + i * w`, where `base` is the relative address of array element `X[0]`, and `w` is the width of each array element. In case of a two-dimensional array, the relative address of the array element $X[i_1][i_2]$ is calculated as $base + i_1 * w_1 + i_2 * w_2$ where w_1 is the width of a row and w_2 is the width of an element in a row. In general, for a k-dimensional array, the formula can be written as follows:

$$base + i_1 * w_1 + i_2 * w_2 + . . . + i_k * w_k \quad (1)$$

We can also determine the relative address of an array element in terms of the number of elements n_k along `k` dimensions of the array with each element of width `w`. In this case, the address calculations are done on the basis of row-major or column-major layout of the arrays. Consider a two-dimensional array `X[2][3]`, which can be stored either in a row-major form or in a column-major form as shown in Figure 7.8.

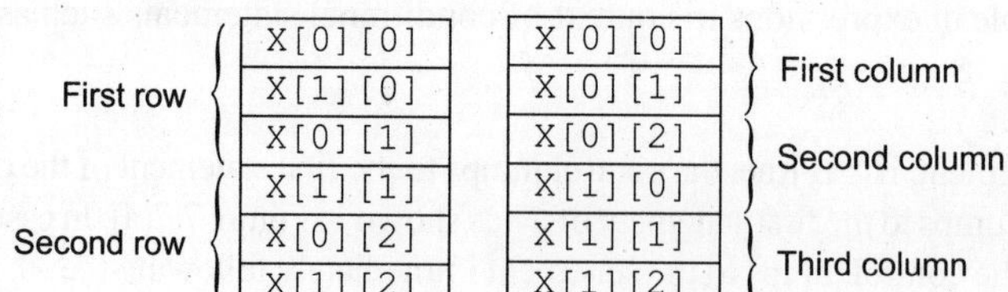

(a) Row-major (b) Column-major

Figure 7.8 Layouts of a Two-dimensional Array

If the elements of a two-dimensional array $X[n_1][n_2]$ are stored in a row-major form, the relative address of an array element $X[i_1][i_2]$ is calculated as follows:

$$base + (i_1 * n_2 + i_2) * w$$

On the other hand, if the elements are stored in a column-major form, the relative address of $X[i_1][i_2]$ is calculated as follows:

$$base + (i_1 + i_2 * n_1) * w$$

The row-major and column-major forms can be generalized to k-dimensions. If we generalize row-major form, then elements are stored in such a way that when we scan down a block of storage, the rightmost scripts appear to vary fastest. On the other hand, in case of column-major form, the leftmost scripts appear to vary fastest.

In general, the array elements in one-dimensional array need not be numbered as `0,1,...,n - 1`, rather they can be numbered as `low, low + 1,..., high`. Now, the address of an array reference `X[i]` can be calculated as `base + (i - low) * w`. Here, `base` is the relative address of `A[low]`.

13. Explain the translation of array references.

Ans: The main problem while translating and generating the intermediate code for an array references is to relate the address calculation formulas to a grammar for array references. Consider the following grammar, where the non-terminal `M` generates an array name followed by a sequence of index expressions:

```
M → M[A] | digit[A]
```

Figure 7.9 shows the translation scheme that generates three-address code for expressions with array references. It comprises the productions and semantic routines for generating three-address code for expressions incrementally. In addition, it comprises the productions involving the non-terminal `M`. We have also assumed that the addresses are calculated using the formula (1), which is based on the width of the array elements.

```
S → digit = A ;    {gen (top.get(digit.lexeme)'=' A.addr);}
   |M = A;         {gen(M.addr.base'['M.addr']''='E.addr);}
A → A1 + A2        {A.addr = newTemp();
                    gen(A.addr'='A1.addr'+'A2.addr);}
   |digit          {A.addr = top.get(digit.lexeme);}
   |M              {A.addr = new Temp();
                    gen(A.addr'='M.array.base'['M.addr']');}
M → digit[A]       {M.array = top.get(digit.lexeme);
                    M.type = M.array.type.elem;
                    M.addr = new Temp();
                    gen(M.addr'='A.addr'*'M.type.width);}
   |M1[A]          {M.array = M1.array;
                    M.type = M1.type.elem;
                    t = new Temp();
                    M.addr = new Temp();
                    gen(t'='A.addr'*'M.type.width);}
                    gen(M.addr'='M1.addr'+'t);}
```

Figure 7.9 Semantic Actions for Array References

In Figure 7.9, the non-terminal `M` has three synthesized attributes `M.addr`, `M.array`, and `M.type`. Here `M.type` represents a temporary to be used during computation of the offsets of i_j * w_j terms, `M.array` denotes a pointer pointing to the symbol table entry for an array name, and `M.type` is the type of the subarray generated by `M`.

In the semantic actions of Figure 7.9, the first production `S → digit = A` shows an assignment to a non-array variable. In the second production `S → M = A`, an index copy instruction is generated by the semantic action to assign the value of expression `A` to the location of array reference `M`. As discussed earlier, the symbol table entry for the array is obtained by the attribute `M.array`. The attribute `M.array.base` gives the base address of the array which is the address of its 0th element and the attribute `M.addr` is a temporary variable that holds the offset for array reference generated by `M`. Thus, `M.array.base [M.addr]` gives the location for the array reference. The r-value from address `A.addr` is copied into the `M`'s location by the generated instruction.

The production `A → M` has a semantic rule associated with it that generates a code which copies value at location `M` into a temporary variable `M.array.base[M.addr]`.

14. What do you understand by procedure calls?

Ans: In high-level languages, we use procedures to break down a large program into smaller modular components, and the procedures that return some value are called **functions**. The procedure or function is an important and frequently used programming construct. A compiler is expected to generate a good three-address code for procedure calls and returns. The runtime routine that handles procedure argument passing, procedure calls and returns, is a part of runtime support package. For example, consider the following grammar for a simple procedure call statement:

```
S → call id(Alist)
Alist → Alist, A
Alist → A
```

The translation of a simple procedure call consists of a calling sequence, which is defined as a sequence of actions performed while entering into and exiting from each procedure.

Calling sequence: On the occurrence of a procedure call, space for the activation record of the called procedure activation record is allocated on the stack. The called procedure arguments are evaluated and are made available to the called procedure by storing them at a known location. The environment pointers must be established so that the called procedure can access data in the enclosing blocks. When a call occurs, the state of the procedure is saved so that the execution of the calling procedure can be resumed after the call. The return address, which is the location to which the called procedure must return after its execution, is also saved at a known location. Finally, a jump to the first statement of the code for the called procedure must be generated.

Return sequence: When the control reaches the last statement of the calling procedure, that is, when the procedure returns, several actions need to be performed. If a procedure call is a function, then the return value (result) must be saved at some known location. The activation record of the calling procedure is then restored and finally a jump to the return address of calling procedure is generated.

Translation of Procedure Calls:

1. `S → call id(Alist),`

```
   {for each item R on queue do
     emit ('param 'R);
   emit('call' id.place)}
```

The code for `S` is the code for `Alist`, which evaluates the arguments, followed by the `param R` statement for each argument, followed by a `call` statement.

2. `Alist → Alist,A`

```
   {append A.place to the end of queue}
```

3. `Alist → A`

```
   {initialize queue to contain only A.place}
```

At this point, the queue is cleared and gets a single pointer to the symbol table location for the name that denotes the value of `A`.

15. Explain declarations with the help of a suitable example.

Ans: The general form of declarations in any programming language is a keyword (that denotes an attribute) followed by a list of names having that attribute. The semantic routine/action associated with each declaration statement should enter the declared attribute into the symbol-table entry for each name on the list. The productions for such declarations are given below:

```
S → double namelist
   |float namelist
namelist→id, namelist
   |id
```

Here, the declarations within productions are in such a form which would raise a problem that if someone wants to get at the attribute associated with the `namelist`, then he has to wait until the entire list of `id`'s has been reduced to `namelist`. Thus, when `id` or `id,namelist` is reduced to `namelist`, the SDT schemes based on these productions cannot create the correct symbol table entries. To avoid this kind of difficulty, we can follow any of these approaches:

- In the first approach, we can create a separate list for `double` and `float` as given below:

```
S → double doublelist
   |float floatlist
doublelist → id, doublelist
   |id
floatlist → id, floatlist
   |id
```

  This approach is based on the assumption that the LR parser would be able to decide whether to reduce the first `id` to `doublelist` or `floatlist`. This approach is not desirable for large numbers of attributes because as the number of attributes increases, the number of productions also increases. This would create a decision problem for the parser.
- In the second approach, we can simply rewrite the grammar rules by considering the translation of names just as a list of names. Now, the above rules can be rewritten as follows:

```
S → S,id
  |double id
  |float id
```

  Now, we can define the semantic actions for these rules as follows:

```
S → double id {Enter(id.place, double); S.attr: = double}
S → float id  {Enter(id.place, float); S.attr: = float}
S → S1, id    {Enter(id.place, S1.attr);
               S.attr: = S1.attr}
```

These semantic actions can now enter the appropriate attribute into the symbol table for each name on the list. Here, `S.attr` is the translation attribute for non-terminal `S`, the procedure `enter(p, x)` associates attribute `x` to the symbol table entry pointed to by `p`, and `id.place` points the symbol table entry for the name represented by the token `id`.

16. Discuss the syntax directed translation of case statements.

Ans: The syntax for `case` or `switch` statement is

```
switch E
begin
case V1: S1
   case V2: S2
   .
   .
   .
   case Vn-1: Sn-1
   default: Sn
end
```

where

E is an expression to be evaluated;

$V_1, V_2, \ldots, V_{n-1}$ are the distinct values and are known as **case labels**, and V_n is a default statement;

$S_1, S_2, \ldots, S_{n-1}$ are the statements that will be executed when a particular value is matched.

The case values are constants and are selected by the selector expression E. First, E is evaluated and the resultant value is matched with these constant values. Then the associated statement sequence of the matched case value is executed. There is a default expression which is always executed if no other value is matched.

Syntax directed translation of case statements: A simple syntax directed translation scheme translates the case statements into an intermediate code as shown in Figure 7.10.

```
                code to evaluate E into t
                goto SAMPLE
        A1:     code for S1
                goto REF
        A2:     code for S2
                goto REF
                .
                .
                .
        An-1:   code for Sn-1
                goto REF
        An:     code for Sn
                goto REF
SAMPLE:         if t = v1 goto A1
                if t = v2 goto A2
                if t = v3 goto A3
                .
                .
                .
                if t = Vn-1 goto An-1
                goto An
REF:
```

Figure 7.10 Translation of Case Statement into Intermediate Code

When the switch keyword is encountered, two labels SAMPLE and REF, and a temporary variable t are generated. As we start parsing, we find the expression E and now we generate a code to evaluate this expression in the temporary t. When E is processed, we generate the jump goto REF.

On the occurrence of each `case` keyword, a new label A_i is created and entered into the symbol table. The cases are stored on a stack, which contains a pointer to this symbol entry along with the value V_i of each case constant. The evaluated expression in temporary `t` is matched with the available values V_1, $V_2, \ldots, V_{n-1}$ and if a value match occurs then the corresponding statements are executed. If no value is matched, then the default case A_n is executed.

Note that all the test expressions appear at the end. This enables a simple code generator to recognize the multiway branch and to generate an efficient code for it. If the branching conditions are placed at the beginning then the compiler would have to perform extensive analysis to generate the most efficient implementation.

17. What is backpatching? Explain.

Ans: The syntax directed definitions can be easily implemented by using two passes. In the first pass, we construct a syntax tree for the input, and in the second pass, we traverse the tree in depth first order to complete the translations in the given definition. Generating code for flow of control statements and Boolean expressions is difficult in single pass. This is because we may not be able to know the labels that the control must goto during the generation of jump statements. Thus, the generated code would be a series of branching statements in which the targets of the jumps are temporarily left unspecified. To overcome this problem, we use back patching, which is a technique to solve the problem of replacing symbolic names into the `goto` statements by the actual target addresses. However, some languages do not permit to use symbolic names in the branches, for this we maintain a list of branches that have the same target labels and then replace them once they are defined. To manipulate the lists of jumps, the following three functions are used:

- **`makelist(i)`** : This function creates a new list containing an index `i` into the array of statements and then returns a pointer pointing to the newly created list.
- **`merge(p1, p2)`** : This function concatenates the two lists pointed to by the pointers p_1 and p_2 respectively, and then returns a pointer to the concatenated list.
- **`backpatch(p, i)`** : This function inserts `i` as the target labels for each of the statements on the list pointed to by `p`.

18. Translate the expression a: = -b * (c + d)/e into quadruples and triple representation.

Ans: The three-address code for the given expression is given below:

```
t1: = -b (here, '-' represents unary minus)
t2: = c + d
t3: = t1 * t2
t4: = t3/e
a:  = t4
```

The quadruple representation of this three-address code is shown in Figure 7.11.

	Operator	Operand 1	Operand 2	Result
0	–	b		t_1
1	+	c	d	t_2
2	*	t_1	t_2	t_3
3	/	t_3	e	t_4
4	=	t_4		a

Figure 7.11 Quadruple Representation for a: = –b * (c + d)/e

The triple representation for the expression is given in Figure 7.12.

	Operator	Operand 1	Operand 2
0	–	b	
1	+	c	d
2	*	(0)	(1)
3	/	(2)	e
4	=	a	(3)

Figure 7.12 Triple Representation for a: = –b * (c + d)/e

19. Translate the expression X = - (a + b) * (c + d) + (a + b + c) into quadruples and triples.

Ans: The three-address code for the given expression is given below:

t_1 : = a + b
t_2 : = $-t_1$
t_3 : = c + d
t_4 : = t_2 * t_3
t_5 : = t_1 + c
t_6 : = t_4 + t_5
X : = t_6

The quadruple representation is shown in Figure 7.13.

	Operator	Operand 1	Operand 2	Result
0	+	a	b	t_1
1	–	t_1		t_2
2	+	c	d	t_3
3	*	t_2	t_3	t_4
4	+	t_1	c	t_5
5	+	t_4	t_5	t_6
6	=	t_6		

Figure 7.13 Quadruple Representation for X = –(a + b) * (c + d) + (a + b + c)

The triple representation for the given expression is shown in Figure 7.14.

	Operator	Operand 1	Operand 2
0	+	a	b
1	–	(0)	
2	+	c	d
3	*	(1)	(2)
4	+	(0)	c
5	+	(3)	(4)
6	=	(5)	

Figure 7.14 Triple Representation for X = –(a + b) * (c + d) + (a + b + c)

20. Generate the three-address code for the following program segment.

```
main()
{
    int k = 1;
    int a[5];
    while (k <= 5)
  {
    a[k] = 0;
    k++;
  }
}
```

Ans: The three-address code for the given program segment is given below:

1. k: = 1
2. if k <= 5 goto (4)
3. goto (8)
4. t_1: = k * width
5. t_2: = addr(a)-width
6. $t_2[t_1]$: = 0
7. t_3: = k + 1
8. k: = t_3
9. goto (2)
10. Next

21. Generate the three-address code for the following program segment

```
while(x < z and y > s) do
if x = 1 then
    z = z + 1
else
    while x <= s do
      x = x + 10;
```

Ans: The three-address code for the given program segment is given below:

1. if x < z goto (3)
2. goto (16)
3. if y > s goto (5)
4. goto (16)
5. if x = 1 goto (7)
6. goto (10)
7. t_1: = z + 1
8. z: = t_1
9. goto (1)

```
10. if x <= s goto (12)
11. goto (1)
12. t2: = x + 10
13. x: = t2
14. goto (10)
15. goto (1)
16. Next
```

22. Consider the following code segment and generate the three-address code for it.

```
for (k = 1; k <= 12; k++)
    if x < y then a = b + c;
```

Ans: The three-address code for the given program segment is given below:

```
1. k: = 1
2. if k <= 12 goto (4)
3. goto (11)
4. if x < y goto (6)
5. goto (8)
6. t1: = b + c
7. a: = t1
8. t2: = k + 1
9. k: = t2
10. goto (2)
11. Next
```

23. Translate the following statement, which alters the flow of control of expressions, and generate the three-address code for it.

```
while(P < Q)do
    if(R < S) then a = b + c;
```

Ans: The three-address code for the given statement is as follows:

```
1. if P < Q goto (3)
2. goto (8)
3. if R < S goto (5)
4. goto (1)
5. t1: = b + c
6. a: = t1
7. goto (1)
8. Next
```

24. Generate the three-address code for the following program segment where, x and y are arrays of size 10 * 10, and there are 4 bytes/word.

```
begin
  add = 0
  a = 1
  b = 1
  do
      begin
        add = add + x[a,b] * y[a,b]
        a = a + 1
        b = b + 1
      end
  while a <= 10 and b <= 10
end
```

Ans: The three-address code for the given program segment is given below:

1. add:= 0
2. a: = 1
3. b: = 1
4. t_1: = a * 10
5. t_1: = t_1 + b
6. t_1: = t_1 * 4
7. t_2: = addr(x) - 44
8. t_3: = t_2[t_1]
9. t_4: = b * 10
10. t_4: = t_4 + a
11. t_4: = t_4 * 4
12. t_5: = addr(y) - 44
13. t_6: = t_5[t_4]
14. t_7: = t_3 * t_6
15. t_7: = add + t_7
16. t_8: = a + 1
17. a: = t_8
18. t_9: = b + 1
19. b: = t_9
20. if a <= 10 goto (22)
21. goto (23)
22. if b <= 10 goto(4)
23. Next

25. Translate the following program segment into three-address statements:

```
switch(a + b)
{
  case 2: {x = y; break;}
  case 5: switch x
          {
            case 0: {a = b + 1; break;}
            case 1: {a = b + 3; break;}
            default: {a = 2; break;}
          }
   break;
  case 9: {x = y - 1; break;}
  default: {a = 2; break;}
}
```

Ans: The three-address code for the given program segment is given below:

1. t_1: = a + b
2. goto (23)
3. x: = y
4. goto (27)
5. goto (14)
6. t_3: = b + 1
7. a: = t_3
8. goto (27)
9. t_4: = b + 3
10. a: = t_4
11. goto (27)
12. a: = 2
13. goto (27)
14. if x = 0 goto (6)
15. if x = 1 goto (9)
16. goto (12)
17. goto (27)
18. t_5: = y - 1
19. a: = t_5
20. goto (27)
21. a: = 2
22. goto (27)
23. if t = 2 goto (3)

```
24. if t = 5 goto (5)
25. if t = 9 goto (18)
26. goto (21)
27. Next
```

Multiple-Choice Questions

1. Which of the following is not true for the intermediate code?
   (a) It can be represented as postfix notation.
   (b) It can be represented as syntax tree, and or a DAG.
   (c) It can be represented as target code.
   (d) It can be represented as three-address code, quadruples, and triples.
2. Which of the following is true for intermediate code generation?
   (a) It is machine dependent.
   (b) It is nearer to the target machine.
   (c) Both (a) and (b)
   (d) None of these
3. Which of the following is true in the context of high-level representation of intermediate languages?
   (a) It is suitable for static type checking.
   (b) It does not depict the natural hierarchical structure of the source program.
   (c) It is nearer to the target program.
   (d) All of these
4. Which of the following is true for the low-level representation of intermediate languages?
   (a) It requires very few efforts by the source program to generate the low-level representation.
   (b) It is appropriate for machine-dependent tasks like register allocation and instruction selection.
   (c) It does not depict the natural hierarchical structure of the source program.
   (d) All of these
5. The reverse polish notation or suffix notation is also known as ———.
   (a) Infix notation
   (b) Prefix notation
   (c) Postfix notation
   (d) None of above
6. In a two-dimensional array `A[i][j]`, where `i` is a element of width w_1 and `j` is of width w_2, the relative address of `A[i][j]` can be calculated by the formula ———.
   (a) $i * w_1 + j * w_2$
   (b) $base + i * w_1 + j * w_2$
   (c) $base + i * w_2 + j * w_1$
   (d) $base + (i + j) * (w_1 + w_2)$

Answers

1. (c) 2. (c) 3. (a) 4. (b) 5. (c) 6. (b)

8

Type Checking

1. What is a type system? List the major functions performed by the type systems.

Ans: A **type system** is a tractable syntactic framework to categorize different phrases according to their behaviors and the kind of values they compute. It uses logical rules to understand the behavior of a program and associates types with each compound value and then it tries to prove that no type errors can occur by analyzing the flow of these values. A type system attempts to guarantee that only value-specific operations (that can match with the type of value used) are performed on the values. For example, the floating-point numbers in C uses floating-point specific operations to be performed over these numbers such as floating-point addition, subtraction, multiplication, etc.

The language design principle ensures that every expression must have a type that is known (at the latest, at run time) and a type system has a set of rules for associating a type to an expression. Type system allows one to determine whether the operators in an expression are appropriately used or not. An implementation of type system is called **type checker**.

There are two type systems, namely, *basic type system* and *constructed type system*.

- **Basic type system:** Basic type system contains atomic types and has no internal structure. They contain integer, real, character, and Boolean. However, in some languages like Pascal, they can have subranges like 1 . . . 10 and enumeration types like orange, green, yellow, amber, etc.
- **Constructed type system:** Constructed type system contains arrays, records, sets, and structure types constructed from basic types and/or from other constructed types. They also include pointers and functions.

Type system provides some functions that include:

- **Safety:** A type system allows a compiler to detect meaningless or invalid code which does not make a sense; by doing this it offers more strong typing safety. For example, an expression `5/"Hi John"` is treated as invalid because arithmetic rules do not specify how to divide an integer by a string.
- **Optimization:** For optimization, a type system can use static and/or dynamic type checking, where static type checking provides useful compile-time information and dynamic type checking verifies and enforces the constraints at runtime.
- **Documentation:** The more expressive type systems can use types as a form of documentation to show the intention of the programmer.

- **Abstraction (or modularity):** Types can help programmers to consider programs as a higher level of representation than bit or byte by hiding lower level implementation.

2. Define type checking. Also explain the rules used for type checking.

Ans: **Type checking** is a process of verifying the type correctness of the input program by using logical rules to check the behavior of a program either at compile time or at runtime. It allows the programmers to limit the types that can be used for semantic aspects of compilation. It assigns types to values and also verifies whether these values are consistent with their definition in the program.

Type checking can also be used for detecting errors in programs. Though errors can be checked dynamically (at runtime) if the target program contains both the type of an element and its value, but a sound type system eliminates the need for dynamic checking for type errors by ensuring that these errors would not arise when the target program runs.

If the rules for type checking are applied strongly (that is, allowing only those automatic type conversions which do not result in loss of information), then the implementation of the language is said to be *strongly typed*; otherwise, it is said to be *weakly typed*. A strongly typed language implementation guarantees that the target program will run without any type errors.

Rules for type checking: Type checking uses syntax-directed definitions to compute the type of the derived object from the types of its syntactical components. It can be in two forms, namely, *type synthesis* and *type inference*.

- **Type synthesis:** Type synthesis is used to build the type of an expression from the types of its sub-expressions. For type synthesis, the names must be declared before they are used. For example, the type of expression $E_1 * E_2$ depends on the types of its sub-expressions E_1 and E_2. A typical rule is used to perform type synthesis and has the following form:

```
if expression f has a type s → t and expression x has a type s,
then expression f(x) will be of type t
```

Here, $s \rightarrow t$ represents a function from s to t. This rule can be applied to all functions with one or more arguments. This rule considers the expression $E_1 * E_2$ as a function, $mul(E_1, E_2)$ and uses E_1 and E_2 to build the type of $E_1 * E_2$.

- **Type inference:** Type inference is the analysis of a program to determine the types of some or all of the expressions from the way they are used. For example,

```
public int mul(int E1, int E2)
    return E1 * E2;
```

Here, E_1 and E_2 are defined as integers. So by type inference, we just need definition of E_1 and E_2. Since the resulting expression $E_1 * E_2$ uses * operation, which would be taken as integer because it is performed on two integers E_1 and E_2. Therefore, the return type of mul must be an integer. A typical rule is used to perform type inference and has the following form:

```
if f(x) is an expression,
then for some type variables α and β, f is of type α → β and
x is of type α
```

3. Explain type expressions.

Ans: **Type expressions** are used to represent structure of types and can be considered as a textual representation for types. A type expression can either be of basic type or can be created by applying an operator (known as **type constructor**) to a type expression.

For example, a type expression for the array type `int[3][5]` considers it as an "array of 3 arrays having 5 integers in each of them", and its type expression can be written as "array (3, array (5, integer))". The type expression uses a tree to represent the type structure. The tree representation of array type `int[3][5]` is shown in Figure 8.1, where an operator array takes two arguments: a number and a type.

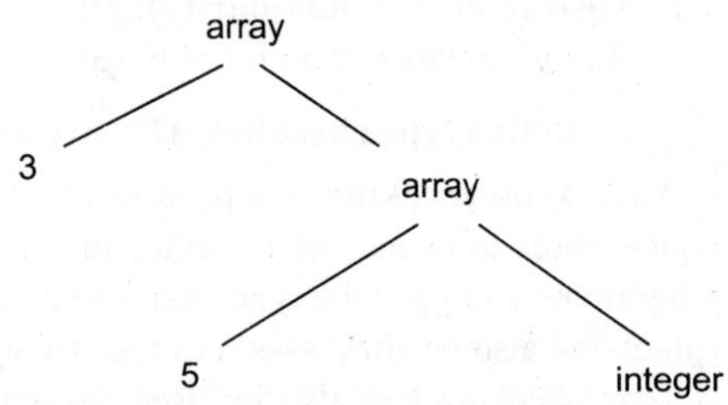

Figure 8.1 Type Expression of Int[3][5]

Type expressions can be defined as follows:

- **Basic types:** Every basic types such as Boolean, char, integer, float, void, etc., is a type expression.
- **Type names:** Every type name is a type expression.
- **Constructed types:** A constructed type applies constructors to the type expressions, which can be:
  - **Arrays:** A type expression can be constructed by applying an array type constructor to a number and a type expression. It can be represented as array `(I,T)`, where `I` is an index type and `T` is the type of array elements. For example, array `(1...10, integer)`.
  - **Cartesian product:** For any type expressions T_1 and T_2, the Cartesian product $T_1 \times T_2$ is also a type expression.
  - **Record with field names:** A record type constructor is applied to the field names to form a type expression. For example, `record{float a, int b}` → X;
  - **Function types:** A type constructor → is used to form a type expression for function types. For example, A → B denotes a function from type A to type B. For example, real × real → real.
  - **Pointers:** A type expression pointer (T_1) represents a pointer to an object of type T_1.

4. Define these terms: static type checking, dynamic type checking, and strong typing.

Ans: Static type checking: In static type checking, most of the properties are verified at compile time before the execution of the program. The languages C, C++, Pascal, Java, Ada, FORTRAN, and many more allow static type checking. It is preferred because of the following reasons:

- As the compiler uses type declarations and determines all types at compile time, hence catches most of the common errors at compile time.
- The execution of output program becomes fast because it does not require any type checking during execution.

The main disadvantage of this method is that it does not provide flexibility to perform type conversions at runtime. Moreover, the static type checking is conservative, that is, it will reject some programs that may behave properly at runtime, but that cannot be statically determined to be well-typed.

Dynamic type checking: Dynamic type checking is performed during the execution of the program and it checks the type at runtime before the operations are performed on data. Some languages that support dynamic type checking are Lisp, Java Script, Smalltalk, PHP, etc. Some advantages of the dynamic type checking are as follows:

- It can determine the type of any data at runtime.
- It gives some freedom to the programmer as it is less concerned about types.
- In dynamic typing, the variables do not have any types associated with them, that is, they can refer to a value of any type at runtime.

- It checks the values of all the data types during execution which results in more robust code.
- It is more flexible and can support union types, where the user can convert one type to another at runtime.

The main disadvantage of this method is that it makes the execution of the program slower by performing repeated type checks.

Strong typing: A type checking which guarantees that no type errors can occur at runtime is called **strong type checking** and the system is called **strongly typed**. The strong typing has certain disadvantages such as:

- There are some checks like array bounds checking which require dynamic checking.
- It can result into performance degradation.
- Generally, these type systems have holes in the type systems, for example, variant records in Pascal.

5. Write down the process to design a type checker.

Ans: A type checker is an implementation of a type system. The process to design a type checker includes the following steps:

1. **Identifying the available types in the language:** We have two types that are available in the language.
   - Base types (integer, double, Boolean, string, and so on)
   - Compound types (arrays, classes, interfaces, and so on)
2. **Identifying the language constructs with associated types:** Each programming language consists of some constructs and each of them is associated with a type as discussed below:
   - **Constants:** Every constant has an associated type. A scanner identifies the types and associated lexemes of a constant.
   - **Variables:** A variable can be global, local, or an instance of a class. Each of these variables must have a declared type, which can either be one of the base types or the supported compound types.
   - **Functions:** The functions have a return type, and the formal parameters in function definition as well as the actual arguments in the function call also have a type associated with them.
   - **Expressions:** An expression can contain a constant, variable, functional call, or some other operators (like unary or binary operators) that can be applied in an expression. Hence, the type of expression depends on the type of constant, variable, operands, function return type, and on the type of operators.
3. **Identifying the language semantic rules:** The production rules to parse variable declarations can be written as:

```
Variable Declaration → Variable
Variable → Type identifier
Type → int|double|Boolean|string|identifier|Type[]
```

The parser stores the name of an identifier lexeme as an attribute attached to the token. The name associated with the identifier symbol, and the type associated with the identifier and type symbol are used to reduce the variable production. A new variable declaration is created by declaring an identifier of that type and that variable is stored in the symbol table for further lookup.

6. What is type equivalence?

Ans: Type equivalence is used by the type checking to check whether the two type expressions are equivalent or not. It can be done by checking the equivalence between the two types. The rule used for type checking works as follows:

```
if two type expressions are equal then return a certain type else
return a type error()
```

When two type expressions are equivalent, we need a precise definition of both the expressions. When names are given to type expressions, and these names are further used in subsequent type expressions, it may result in potential ambiguities. The key issue is whether a name in a type expression stands for itself, or it is an abbreviation for another type expression.

There are two schemes to check type equivalence of expressions:

Structural equivalence: Structural equivalence needs a graph to represent the type expressions. The two type expressions are said to be structurally equivalent if and only if they hold any of the following conditions:

- They are of identical basic type.
- Same type constructor has been applied to equivalent types to construct the type expressions.
- A type name of one represents the other.

Name equivalence: If type names are treated as standing for themselves, then the first two conditions of structural equivalence lead to another equivalence of type expressions called name equivalence. In other words, name equivalence considers types to be equal if and only if the same type names are used and one of the first two conditions of structure equivalence holds.

For example, consider the following few types and variable declarations.

```
typedef double Value
. . .
. . .
Value var1, var2
Sum var3, var4
```

In these statements, `var1` and `var2` are name equivalent, so are `var3` and `var4`, because their type names are same. However, `var1` and `var3` are not name equivalent, because their type names are different.

7. Explain type conversion.

Ans: Type conversion refers to the conversion of a certain type into another by using some semantic rules. Consider an expression `a + b`, where `a` is of type `int` and `b` is of `float`. The representations of floats and integers are different within a computer, and an operation on integers and floats uses different machine instructions. Now, the primary task of the compiler is to convert one of the operands of + to make both of the operands to same type. For example, an expression `5 * 7.14` has two types, one is of `float` type and other one is of type `int`. To convert integer type constant into `float` type, we use a unary operator (`float`) as shown here:

```
x = (float) 5
y = x * 7.14
```

The type conversion can be done implicitly or explicitly. The conversion from one type to another is called **implicit**, if it is automatically done by the computer. Usually, implicit conversions of constants can be done at compile time and it results in an improvement in the execution time of the object program. Implicit type conversion is also known as **coercion**. A conversion is said to be **explicit** if the programmer must write something to cause the conversion. For example, all conversions in Ada are explicit. Explicit conversions can be considered as a function applications to a type checker. Explicit conversion is also known as **casts**.

Conversion in languages can be considered as *widening conversions* and *narrowing conversions*, as shown in Figure 8.2(a) and (b), respectively.

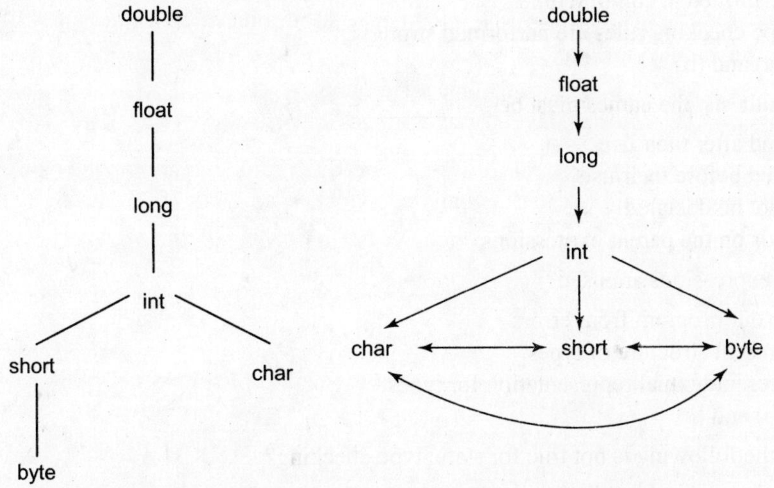

(a) Widening conversion (b) Narrowing conversion

Figure 8.2 Conversion Between Primitive Types in Java

The rules used for widening is given by the hierarchy in Figure 8.2(a) and are used to preserve the information. In widening hierarchy, any lower type can be widened into a higher type like a `byte` can be widened to a `short` or to an `int` or to a `float,` but a `short` cannot be widened to a `char.`

The narrowing conversions, on the other hand, may result in loss of information. The rules used for narrowing is given by the hierarchy in Figure 8.2(b), in which a type x can be narrowed to type y if and only if there exists a path from x to y. Note that `char`, `short,` and `byte` are pairwise convertible to each other.

Multiple-Choice Questions

1. Which of the following is true for type system?
   (a) It is a tractable syntactic framework.
   (b) It uses logical rules to determine the behavior of a program.
   (c) It guarantees that only value specific operations are allowed.
   (d) All of these
2. A type system can be __________ type system or __________ type system.
   (a) Basic, constructed (b) Static, dynamic
   (c) Simple, compound (d) None of these

3. Which of the following is true for type checking?
   (a) It ensures type correctness.
   (b) It can only be done at compile time.
   (c) It can only be done at runtime.
   (d) All of these
4. A type checking is called strongly typed if __________.
   (a) It is performed at runtime.
   (b) It is performed at compile time.
   (c) The type checking rules are performed strongly.
   (d) Both (a) and (b)
5. In type synthesis, the names must be __________.
   (a) Declared after their use
   (b) Declared before their use
   (c) Need not be declared
   (d) Depends on the parent expressions
6. Why type expressions are used?
   (a) To free our program from errors
   (b) To represent structure of types
   (c) To represent textual representation for types
   (d) Both (b) and (c)
7. Which of the following is not true for static type checking?
   (a) It is performed at compile time.
   (b) It catches errors at compile time.
   (c) Most of the properties are verified at compile time.
   (d) It provides flexibility of performing type conversions at runtime.
8. A strong type checking ensures that __________.
   (a) No type errors can occur at compile time.
   (b) No type errors can occur at runtime.
   (c) Both (a) and (b)
   (d) None of these
9. Implicit type checking is also known as __________.
   (a) Casts
   (b) Explicit conversion
   (c) Manual conversion
   (d) Coercion

Answers

1. (a) 2. (a) 3. (a) 4. (c) 5. (b) 6. (d) 7. (d) 8. (b) 9. (d)

9

Runtime Administration

1. Define runtime environment. What are the issues in runtime environment?

Ans: The source language definition contains various abstractions such as names, data types, scopes, bindings, operators, parameters, procedures, and flow of control constructs. These abstractions must be implemented by the compiler. To implement these abstractions on target machine, compiler needs to cooperate with the operating system and other system software. For the successful execution of the program, compiler needs to create and manage a **runtime environment**, which broadly describes all the runtime settings for the execution of programs.

In case of compilation of a program, the runtime environment is indirectly controlled by generating the code to maintain it. However, in case of interpretation of the program, the runtime environment is directly maintained by the data structures of the interpreter.

The runtime environment deals with several issues which are as follows:

- The allocation and layout of storage locations for the objects used in the source program.
- The mechanisms for accessing the variables used by the target program.
- The linkages among procedures.
- The parameter passing mechanism.
- The interface to input/output devices, operating systems and other programs.

2. What are the important elements in runtime environment?

Ans: The important elements that constitute a runtime environment for a program are as follows:

- **Memory organization:** During execution, a program requires certain amount of memory for storing the local and global variables, source code, certain data structures, and so on. The way memory is organized for storing these elements is an important characteristic of runtime environment. Different programming languages support different memory organization schemes. For example, `C++` supports the use of pointers and dynamic memory with the help of `new()` and `delete()` functions, whereas FORTRAN 77 does not support pointers and usage of dynamic memory.
- **Activation records:** The execution of a procedure in a program is known as the **activation** of the procedure. The activation of procedures or functions is managed with the help of a contiguous

block of memory known as **activation record**. Activation record can be created statically or dynamically. Statically, a single activation record can be constructed, which is common for any number of activations. Dynamically, number of activation records can be constructed, one for each activation. The activation record contains the memory for all the local variables of the procedure, depending on the way by which activation record is created, the target code has to be generated accordingly to access the local variables.

- **Procedure calling and return sequence:** Whenever a procedure is invoked or called, certain sequence of operations need to be performed, which include evaluation of function arguments, storing it at a specified memory location, transferring the control to the called procedure, etc. This sequence of operations is known as **calling sequence** and **procedure calling**. Similarly, when the activated procedure terminates, some other operations need to be performed such as fetching the return value from a specified memory location, transferring the control back to the calling procedure, etc. This sequence of operations is known as **return sequence**. The calling sequence and return sequence differ from one language to another, and in some cases even from one compiler to another for the same language.
- **Parameter passing:** The functions used in a program may accept one or more parameters. The values of these parameters may or may not be modified inside the function definition. Moreover, the modified values may or may not be reflected in the calling procedure depending on the language used. In some languages like PASCAL and C++, some rules are specified which determine whether the modified value should be reflected in the calling procedure. In certain languages like FORTRAN77 the modified values are always reflected in the calling procedure. There are several techniques by which parameters can be passed to functions. Depending on the parameter passing technique used, the target code has to be generated.

3. Give subdivision of runtime memory.

Or

What is storage organization?

Or

Explain stack allocation and heap allocation?

Or

What is dynamic allocation? Explain the techniques used for dynamic allocation (stack and heap allocation).

Ans: The target program (already compiled) is executed in the runtime environment within its own logical address space known as **runtime memory**, which has a storage location for each program value. The compiler, operating system, and the target machine share the organization and management of this logical address. The runtime representation of the target program in the logical address space comprises data and program areas as shown in Figure 9.1. These areas consist of the following information:

- The generated target code
- Data objects
- Information to keep track of procedure activations.

Since the size of the target code is fixed at compile time, it can be placed in a statically determined area named *Code* (see Figure 9.1), which is usually placed in the low end of memory. Similarly, the memory occupied by some program data objects such as global constants can also be determined at compile time. Hence, the compiler can place them in another statically determined area of memory named *Static*. The main reason behind the static allocation of as many data objects as possible is that the compiler could compile the addresses of these objects into the target code. For example, all the data objects in FORTRAN are statically allocated.

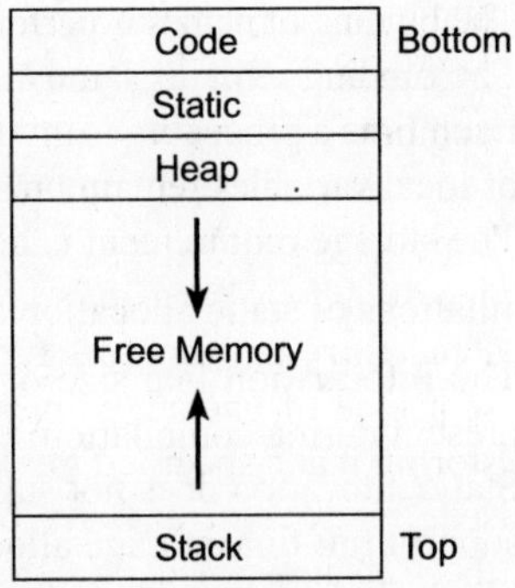

Figure 9.1 Subdivision of Runtime Memory

The other two areas, namely, *Stack* and *Heap* are used to maximize the utilization of space of runtime. The size of these areas is not fixed, that is, as the program executes their size can change. Hence, these areas are dynamic in nature.

Stack allocation: The stack (also known as **control stack** or **runtime stack**) is used to store activation records that are generated during procedure calls. Whenever a procedure is invoked, the activation record corresponds to that procedure is pushed onto the stack and all local items of the procedure are stored in the activation record. When the execution of procedure is completed, the corresponding activation record is popped from the stack and the values of locals are deleted.

The stack is used to manage and allocate storage for the active procedure such that

- On the occurrence of a procedure call, the execution of the calling procedure is interrupted, and the activation record for the called procedure is constructed. This activation record stores the information about the status of the machine.
- On receiving control from the procedure call, the values in the relevant registers are restored and the suspended activation of the calling procedure is resumed, and then the program counter is updated to the point immediately after the call. The stack area of runtime storage is used to store all this information.
- Some data objects which are contained in this activation and their relevant information are also stored in the stack.
- The size of the stack is not fixed. It can be increased or decreased according to the requirement during program execution.

Runtime stack is used in C and Pascal.

Heap allocation: The main limitation of stack area is that it is not possible to retain the values of non-local variables even after the activation record. This is because of last-in-first-out nature of stack allocation. To retain the values of such local variables, heap allocation is used. The heap allocates a contiguous memory locations as and when required for storing the activation records and other data elements. When the activation ends, the memory is deallocated, and this free space can be further used by the heap manager. The heap management can be made efficient by creating a linked list of free blocks. Whenever some memory is deallocated, the free block is appended in the linked list, and when memory needs to be allocated, the most suitable (best-fit) memory block is used for allocation. The heap manager dynamically allocates the memory, which results into a runtime overhead of taking care of defragmentation and garbage collection. The garbage collection enables the runtime system to automatically detect unused data elements and reuse their storage.

4. Explain static allocation. What are its limitations?

Ans: An allocation is said to be static if all data objects are stored at compile time. It has the following properties:

- The binding of names is performed during compilation and no runtime support package is required.
- The binding remains same at runtime as well as compile time.
- Each time a procedure is invoked, the names are bounded to the same storage locations. The values of local variables remain unchanged before and after the transfer of controls.
- The storage requirement is determined by the type of a name.

Limitations of static allocation are given as follows:

- The information like size of data objects and constraints on their memory position needs to be present during compilation.
- Static allocation does not support any dynamic data structures, because there is no mechanism to support run-time storage allocation.
- Since all the activations of a given procedure use the same bindings for local names, recursion is not possible in static allocation.

5. Explain in brief about control stack.

Ans: A stack representing procedure calls, return, and flow of control is called a **control stack** or **runtime stack**. Control stack manages and keeps track of the activations that are currently in progress. When the activation begins, the corresponding activation node is pushed onto the stack and popped out when the activation ends. The control stack can be nested as the procedure calls or activations nest in time such that if `p` calls `q`, then the activation of `q` is nested within the activation of `p`.

6. Define activation tree.

Ans: During the execution of program, activation of the procedures can be represented by a tree known as **activation tree**. It is used to depict the flow of control between the activations. Activations are represented by the nodes in activation tree where each node corresponds to one activation, and the root node represents the activation of the `main` procedure that initiates the program execution. Figure 9.2 shows that the `main()` activates two procedures P_1 & P_2. The activations of procedures P_1 & P_2 are represented in the order in which they are called, that is, from left to right. It is important to note that the left child node must finish its execution before the activation of right node can begin. The activation of P_2 further activates two procedures P_3 and P_4. The flow of control between the activations can be depicted by performing a depth first traversal of the activation tree. We start with the root of the tree. Each node is visited before its child nodes are visited and the child nodes are visited from left to right. When all the child nodes of a particular node have been visited, we can say that the procedure activation corresponding to a node is completed.

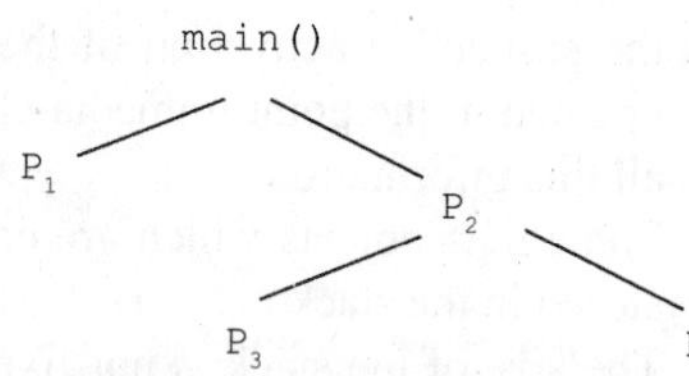

Figure 9.2 Activation Tree

7. Discuss in detail about activation records.

Ans: The **activation record** is a block of memory on the control stack used to manage information for every single execution of a procedure. Each activation has its own activation record with the root of activation tree at the bottom. The path from one activation to another in the activation tree is determined by the corresponding sequence of activation records on the control stack.

Different languages have different activation record contents. In FORTRAN, the activation records are stored in the static data area while in C and Pascal, the activation records are stored in stack area. The contents of activation records are shown in Figure 9.3.

The various fields of activation record are as follows:

Actual parameters
Returned values
Control link (Dynamic link)
Access link (Static link)
Saved machine status
Local data (variables)
Temporaries

Figure 9.3 Activation Record Model

- **Temporaries:** The temporaries are used to store intermediate results that are generated during the evaluation of an expression and cannot be held in registers.
- **Local data:** This field contains local data like local variables, which are local to the execution of a procedure stored in the activation record.
- **Saved machine status:** This field contains the information regarding the state of a machine just before the procedure is called. This information consists of the machine register contents and the return address (program counter value).
- **Access link:** It is an optional field and also called **static link** field. It is a link to non-local data in some other activation record which is needed by the called procedure.
- **Control link:** It is also an optional field and is called **dynamic link** field. It points to the activation record of the calling procedure.
- **Returned value:** It is also an optional field. It is not necessary that all the procedures return a value, but if the procedure does, then for better efficiency this field stores the return value of the called procedure.
- **Actual parameters:** This field contains the information about actual parameters which are used by the calling procedure.

8. Explain register allocation.

Ans: On the target machine, registers are the fastest for the computation as fetching the data from the registers is easy and efficient as compared to fetching it from the memory. So the instructions that involve the use of registers are much smaller and faster than those using memory operands. The main problem with the usage of the register is that the system has limited number of registers which are not enough to hold all the variables. Thus, an efficient utilization of registers is very important to generate a good code. The problem of using registers is divided into two sub problems:

- **Register allocation:** It includes selecting a set of variables to be stored within the registers during program execution.
- **Register assignment:** It includes selecting a specific register to store a variable.

The purpose of register allocation is to map a large number of variables into a few numbers of registers, which may result in sharing of single register by several variables. However, since two variables in use cannot be kept in the same register at the same time, therefore, the variables that cannot be assigned to any register must be kept in the memory.

Register allocation can be done either in intermediate language or in machine language. If register allocation is done during intermediate language, then the same register allocator can be used for several target machines. Machine languages, on the other hand, initially use symbolic names for registers, and register allocation turns these symbolic names into register numbers.

It is really difficult to find an optimal assignment of registers. Mathematically, the problem to find a suitable assignment of registers can be considered as a NP-Complete problem. Sometimes, the target machine's operating system or hardware uses certain register-usage convention to be observed, which makes the assignment of registers more difficult.

For example, in case of integer division and integer multiplication, some machines use even/odd register pairs to store operands and the results. The general form of a multiplication instruction is as follows:

```
M a,b
```

Here, operand a is a multiplicand, and is in the odd register of an even/odd register pair and b is the multiplier, and it can be stored anywhere. After multiplication, the entire even/odd register pair is occupied by the product.

The division instruction is written as

```
D a,b
```

Here, dividend a is stored in the even register of an even/odd register pair and the divisor b can be stored anywhere. After division, remainder is stored in the even register and quotient is stored in the odd register.

Now, consider the two three-address code sequences given in Figure 9.4(a) and (b).

These three-address code sequences are almost same; the only difference is the operator in the second statement. The assembly code sequences for these three-address code sequences are given in Figure 9.5(a) and (b).

Here, L, ST, S, M, and D stand for load, store, subtract, multiply, and divide respectively. R_0 and R_1 are machine registers and SRDA stands for Shift-Right-Double-Arithmetic. SRDA R_0, 32 shifts the dividend into R_1 and clears R_0 to make all bits equal to its sign bit.

```
x = a - b
x = x * c
x = x/d
```

(a)

```
x = a - b
x = x - c
x = x/d
```

(b)

Figure 9.4 Two Three-address Code Sequences

```
L    R1, a
S    R1, b
M    R0, c
D    R0, d
ST   R1, x
```

(a)

```
L    R0, a
S    R0, b
S    R0, c
SRDA R0, 32
D    R0, d
ST   R1, x
```

(b)

Figure 9.5 Assembly Code (Machine Code) Sequences

9. Explain the various parameter passing mechanisms of a high-level language.

Or

What are the various ways to pass parameters in a function?

Ans: When one procedure calls another, the communication between the procedures occurs through non-local names and through parameters of the called procedure. All the programming languages have two types of parameters, namely, *actual parameters* and *formal parameters*. The **actual parameters** are those parameters which are used in the call of a procedure; however, **formal parameters** are those which are used during the procedure definition. There are various parameter passing methods but most of the recent programming languages use *call by value* or *call by reference* or both. However, some older programming languages also use another method *call by name*.

- **Call by value:** It is the simplest and most commonly used method of parameter passing. The actual parameters are evaluated (if expression) or copied (if variable) and then their r-values are passed to the called procedure. **r-value** refers to the value contained in the storage. The values of actual parameters are placed in the locations which belong to the corresponding formal parameters of the called procedure. Since the formal and actual parameters are stored in different memory locations, and formal parameters are local to the called procedure, the changes made in the values of formal parameters are not reflected in the actual parameters. The languages C, C++, Java, and many more use call by value method for passing parameters to the procedures.
- **Call by reference:** In call by reference method, parameters are passed by reference (also known as **call by address** or **call by location**). The caller passes a pointer to the called procedure, which points to the storage address of each actual parameter. If the actual parameter is a name or an expression having an l-value, then the l-value itself is passed (here, l-value represents the address of the actual parameter). However, if the actual parameter is an expression like a + b or 2, having

no l-value, then that expression is calculated in a new location, and the address of that new location is passed. Thus, the changes made in the calling procedure are reflected in the called procedure.

- **Call by name:** It is a traditional approach and was used in early programming languages, such as ALGOL 60. In this approach, the procedure is considered as a macro, and the body of the procedure is substituted for the call in the caller and the formals are literally substituted by the actual parameters. This literal substitution is called **macro expansion** or **in-line expansion.** The names of the calling procedure are kept distinct from the local names of the called procedure. That is, each local name of the called procedure is systematically renamed into a distinct new name before the macro expansion is done. If necessary, the actual parameters are surrounded by parentheses to maintain their integrity.

10. What is the output of this program, if compiler uses following parameter passing methods?

(1) Call by value (2) Call by reference (3) Call by name

The program is given as:

```
void main (void)
{
    int a, b;
    void A(int, int, int);
    a = 2, b = 3;
    A(a + b, a, a);
    printf ("%d", a);
}
void A (int x, int y, int z)
{
    y = y + 1;
    z = z + x;
}
```

Ans: Call by value: In call by value, the actual values are passed. The values `a = 2` and `b = 3` are passed to the function `A` as follows:

```
A (2 + 3, 2, 2);
```

The value of `a` is printed as `2`, because the updated value is not reflected in `main()`.

Call by reference: In call by reference, both formal parameters `y` and `z` have the same reference that is, `a`. Thus, in function `A` the following values are passed.

```
x = 5
y = 2
z = 2
```

After the execution of `y = y + 1`, the value of `y` becomes

```
y = 2 + 1 = 3
```

Since `y` and `z` are referring to the same memory location, `z` also becomes `3`. Now after the execution of statement `z = z + x`, the value of `z` becomes

```
z = 3 + 5 = 8
```

When control returns to `main()`, the value of `a` will now become `8`. Hence, output will be `8`.

Call by name: In this method, the procedure is treated as macro. So, after the execution of the function

```
x = 5
y = y + 1 = 2 + 1 = 3
z = z + x = 2 + 5 = 7
```

When control returns to `main()`, the value of `a` becomes `7`. Hence, output will be `7`.

Multiple-Choice Questions

1. What are the issues that the runtime environment deals with?
   (a) The linkages among procedures
   (b) The parameter passing mechanism
   (c) Both (a) and (b)
   (d) None of these
2. The elements of runtime environment include __________.
   (a) Memory organization
   (b) Activation records
   (c) Procedure calling, return sequences, and parameter passing
   (d) All of these
3. Which of the following area in the memory is used to store activation records that are generated during procedure calls?
   (a) Heap
   (b) Runtime stack
   (c) Both (a) and (b)
   (d) None of these
4. __________ are used to depict the flow of control between the activations of procedures.
   (a) Binary trees
   (b) Data flow diagrams
   (c) Activation trees
   (d) Transition diagram
5. The __________ is a block of memory on the control stack used to manage information for every single execution of a procedure.
   (a) Procedure control block
   (b) Activation record
   (c) Activation tree
   (d) None of these
6. __________ is the process of selecting a set of variables that will reside in CPU registers.
   (a) Register allocation
   (b) Register assignment
   (c) Instruction selection
   (d) Variable selection

7. Which of the following is the parameter passing mechanism of a high-level language?
   (a) Call by value
   (b) Call by reference
   (c) Both (a) and (b)
   (d) None of these

Answers

1. (c) 2. (d) 3. (b) 4. (c) 5. (b) 6. (a) 7. (c)

10

Symbol Table

1. What is symbol table and what kind of information it stores? Discuss its capabilities and also explain the uses of symbol table.

Ans: A **symbol table** is a compile time data structure that is used by the compiler to collect and use information about the source program constructs, such as variables, constants, functions, etc. The symbol table helps the compiler in determining and verifying the semantics of given source program. The information in the symbol table is entered in the lexical analysis and syntax analysis phase, however, is used in later phases of compiler (semantic analysis, intermediate code generation, code optimization, and code generation). Intuitively, a symbol table maps names into declarations (called attributes), for example, mapping a variable name `a` to its data type `char`.

Each time a name is encountered in the source program, the compiler searches it in the symbol table. If the compiler finds a new name or new information about an existing name, it modifies the symbol table. Thus, an efficient mechanism must be provided for retrieving the information stored in the table as well as for adding new information to the table. The entries in the symbol table consists of (name, information) pair. For example, for the following variable declaration statement,

```
char a;
```

The symbol table entry contains the name of the variable along with its data type.

More specifically, the symbol table contains the following information:

- The character string (or lexeme) for the name. If the same name is assigned to two or more identifiers which are used in different blocks or procedures, then an identification of the block or procedure to which this name belongs to must also be stored in the symbol table.
- For each type name, the type definition is stored in the symbol table.
- For each variable name, its type (`int`, `char`, or `real`), its form (label, simple variable, or array), and its location in the memory must also be stored. If the variable is an array, then some other attributes such as its dimensions, and its upper and lower limits along each dimension are also stored. Other attributes such as storage class specifier, offset in activation record, etc. can also be stored.
- For each function and procedure, the symbol table contains its formal parameter list and its return type.
- For each formal parameter, its name, type and type of passing (by value or by reference) is also stored.

A symbol table must have the following capabilities:

- **Lookup:** To determine whether a given name is in the table.
- **Insert:** To add a new name (a new entry) to the table.
- **Access:** To access the information related with the given name.
- **Modify:** To add new information about a known name.
- **Delete:** To delete a name or group of names from the table.

The information stored in the symbol table can be used during several stages of compilation process as discussed below:

- In semantic analysis, it is used for checking the usage of names that are consistent with respect to their implicit and explicit declaration.
- During code generation, it can be used for determining how much and what kind of runtime storage must be allocated to a name.
- The information in the symbol table also helps in error detection and recovery. For example, we can determine whether a particular error message has been displayed before, and if already displayed then avoid displaying it again.

2. What are the symbol table requirements? What are the demerits in the uniform structure of symbol table?

Ans: The basic requirements of a symbol table are as follows:

- **Structural flexibility:** Based on the usage of identifier, the symbol table entries must contain all the necessary information.
- **Fast lookup/search:** The table lookup/search depends on the implementation of the symbol table and the speed of the search should be as fast as possible.
- **Efficient utilization of space:** The symbol must be able to grow or shrink dynamically for an efficient usage of space.
- **Ability to handle language characteristics:** The characteristic of a language such as scoping and implicit declaration needs to be handled.

Demerits in Uniform Structure of Symbol Table:

- The uniform structure cannot handle a name whose length exceed upper bound or limit of name field.
- If the length of a name is small, then the remaining space is wasted.

3. Write down the operations performed on a symbol table.

Ans: The following operations can be performed on a symbol table:

- **Insert:** The insert operation inserts a new name into the table and returns an index of new entry. The syntax of `insert` function is as follows:

```
insert(String key, Object binding)
```

  For example, the function `insert(s,t)` inserts a new string `s` in the table and returns an index of new entry for string `s` and token `t`.

- **Lookup**: This operation searches the symbol table for a given name. The syntax of `lookup` function is as follows:

```
object_lookup(string key)
```

For example, the function `object_lookup(s)` returns an index of the entry for the string `s`; if `s` is not found, it returns 0.

- **Search/Insert:** This operation searches for a given name in the symbol table, and if not found, it inserts it into the table.
- **begin_scope () and end_scope ():** The `begin_scope()` begins a new scope, when a new block starts, that is, when the token { is encountered. The `end_scope()` removes the scope when the scope terminates, that is, when the token } is encountered. After removing a scope, all the declarations inside this scope are also removed.
- **Handling reserved keywords:** Reserved keywords like 'PLUS', 'MINUS', 'MUL', etc., are handled by the symbol table in the following manner.

```
insert ("PLUS", PLUS);
insert ("MINUS", MINUS);
insert ("MUL", MUL);
```

The first 'PLUS', 'MINUS', and 'MUL' in the `insert` operation indicate lexeme and second one indicate the token.

4. Explain symbol table implementation.

Ans: The implementation of a symbol table needs a particular data structure, depending upon the symbol table specifications. Figure 10.1 shows the data structure for implementation of a symbol table.

The character string forming an identifier is stored in a separate array `arr_lexeme`. Each string is terminated by an EOS (end of string character), which is not a part of identifiers. Each entry in the symbol table `arr_symbol_table` is a record having two or more fields, where first field named `lexeme_pointer`

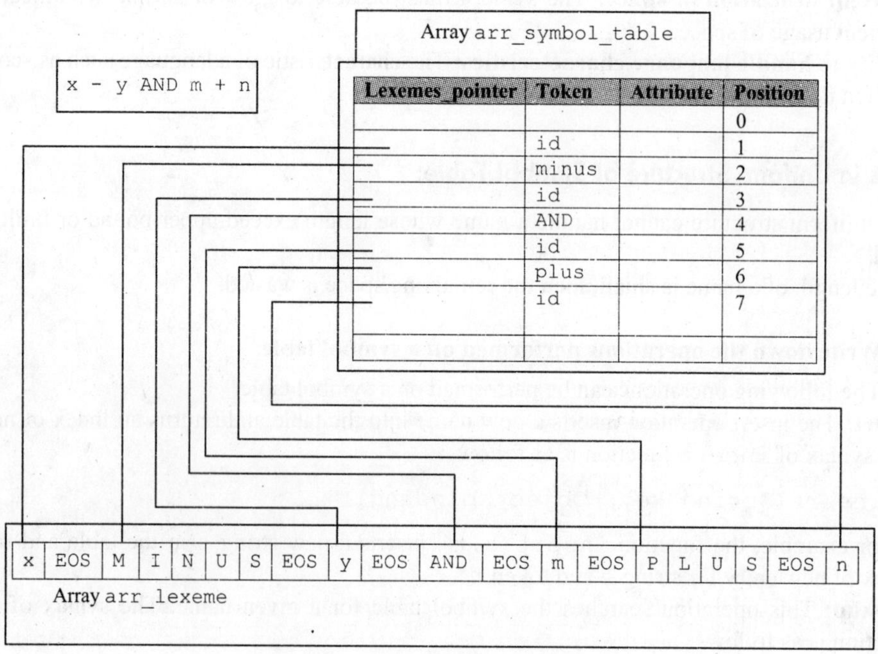

Figure 10.1 Implementation of Symbol Table

points to the beginning of the lexeme, and the second field `Token` consists of the token name. Symbol table also contains two more fields, namely `attribute`, which holds the attribute values, and `position`, which indicates the position of a lexeme in the symbol table are used.

Note that the 0th entry in the symbol table is left empty, as a lookup operation returns 0, if the symbol table does not have an entry for a particular string. The 1st, 3rd, 5th and 7th entries are for the `x`, `y`, `m`, and `n` respectively. The 2nd, 4th and 6th entries are reserved keyword entries for `MINUS`, `AND` and `PLUS` respectively.

Whenever the lexical analyzer encounters a letter in an input string, it starts storing the subsequent letters or digits in a buffer named `lex_buffer`. It then scans the symbol table using the `object_lookup()` operation to determine whether the collected string is in the symbol table. If the lookup operation returns 0, that is, there is no entry for the string in `lex_buffer`, a new entry for the identifier is created using `insert()`. After the insert operation, the index `n` of symbol table entry for the entered string is passed to the parser by setting the `tokenval` to `n`, and an entry in the `Token` field of the token is returned.

5. Discuss various approaches used for organization of symbol table.

Or

Explain the various data structure used for implementing the symbol table.

Ans: The various data structures used for implementing the symbol table are linear list, self-organizing list, hash table, and search tree. The organization of symbol table depends on the selection of the data structure scheme used to implement the symbol table. The data structure schemes are evaluated on the basis of access time, simplicity, storage and performance.

- **Linear list:** A linear list of records is the simplest data structure and it is easiest-to-implement data structure as compared to other data structures for organizing a symbol table. A single array or collection of several arrays is used to store names and their associated information. It uses a simple linear linked list to arrange the names sequentially in the memory.

  The new names are added to the table in the order of their arrival. Whenever a new name is added, the whole table is searched linearly or sequentially to check whether the name is already present in the symbol table or not. If not, then a record for the new name is created and added to the linear list at a location pointed to by the `space` pointer, and the pointer is incremented to point to the next empty location (See Figure 10.2).

Variable	Information (type)	Space (byte)
a	int	2
b	char	1
c	float	4
d	long	4

Space

Figure 10.2 Symbol Table as a Linear List

To access a particular name, the whole table is searched sequentially from its beginning until it is found. For a symbol table having `n` entries, it will take on average `n/2` comparisons to find a particular name.

- **Self-organizing list:** We can reduce the time of searching the symbol table at the cost of a little extra space by adding an additional `LINK` field to each record or to each array index. Now, we search the list in the order indicated by links. A new name is inserted at a location pointed to by `space` pointer, and then all other existing links are managed accordingly. A self-organizing list is shown in Figure 10.3, where the attributes `id1` is related to `id2` and `id3` is related to `id1`, and are linked by the `LINK` pointer.

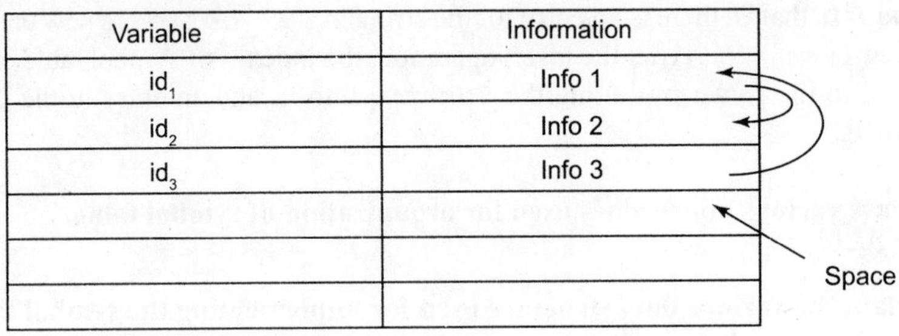

Figure 10.3 Symbol Table as Self Organizing List

The main reason for using the self-organizing list is that if a small set of names is heavily used in a section of program, then these names can be placed at the top while that section is being processed by the compiler. However, if references are random, then the self-organizing list will cost more time and space.

Demerits of self-organizing list are as follows:

- It is difficult to maintain the list if a large set of names is frequently used.
- It occupies more memory as it has a `LINK` field for each record.
- As self-organizing list organizes it itself, so it may cause problems in pointer movements.

- **Hash Table:** A hash table is a data structure that associates keys with values. The basic hashing scheme has two parts:
  - A hash table consisting of a fixed array of `k` pointers to table entries.
  - A storage table with the table entries organized into `k` separate linked lists and each record in the symbol table appears on exactly one of these lists.

To add a name in the symbol table, we need to determine the hash value of that name with the help of a hash function, which maps the name to the symbol table by assigning an integer between `0` to `k - 1`. To search a given name into the symbol, a hash function is applied to that name. Thus, we need to search only that list to determine whether that name exists in the symbol table or not. There is no need to search the entire symbol table. If the name is not present in the list, we create a new record for that name and then insert that record at the head of the list whose index is computed by applying the hash function to the name.

A hash function should be chosen in such a way that it distributes the names uniformly among the `k` lists, and it can be computed easily for the names comprising character strings. The main advantage of using hash table is that, we can insert or delete any name in O(n) time and search any name in O(1) time. However, in the worst case it can be as bad as O(n).

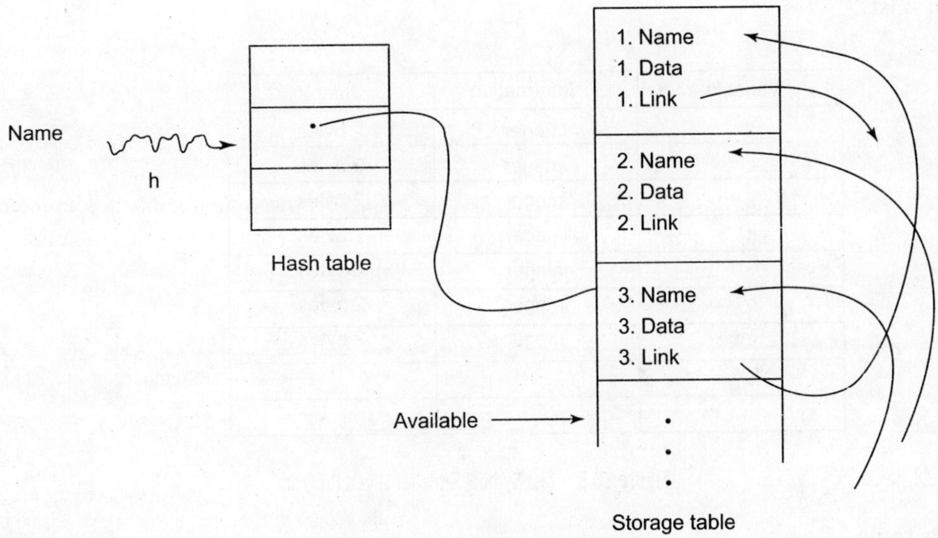

Figure 10.4 Symbol Table as Hash Table

- **Search Tree:** Search tree is an approach to organize symbol table by adding two link fields, `LEFT` and `RIGHT`, to each record. These two fields are used to link the records into a binary search tree. All names are created as child nodes of root node that always follow the properties of a binary search tree.
  - The name in each node is a key value, that is, no two nodes can have identical names.
  - The names in the nodes of left sub tree, if exists, is smaller than the value in the root node.
  - The names in the nodes of right sub tree, if exists, is greater than the value in the root node.
  - The left and right sub trees, if exists, are also binary search trees.

  For example: `name` < `name_i` and `name_i` < `name`. These two statements show that all `name` smaller than `name_i` must be left child of `name_i`; and all `name` greater than `name_i` must be right child of `name_i`. To insert, search and delete in search tree, the binary search tree insert, search and deletion algorithms are followed respectively.

6. Create list, search tree and hash table for given program.

```
int i, j, k;
int mul (int a, int b)
{
    i = a * b;
    Return (i)
}
main ()
{
    int x;
    x = mul (2, 3);
}
```

Ans: List

Variable	Information	Space
x	integer	2 bytes
i	integer	2 bytes
j	integer	2 bytes
k	integer	2 bytes
a	integer	2 bytes
b	integer	2 bytes
mul	integer	2 bytes

← Space

Figure 10.5 List Symbol Table for Given Program

Hash Table

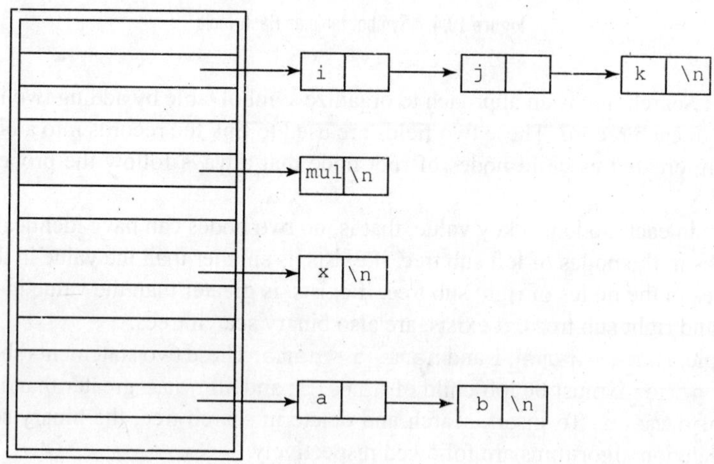

Figure 10.6 Hash Symbol Table for Given Program

Search Tree

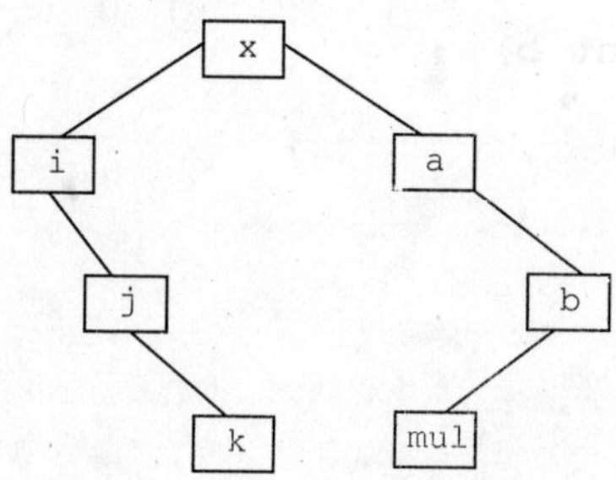

Figure 10.7 Search Tree Symbol Table for the Given Program

7. Discuss how the scope information is represented in a symbol table.

Ans: Scope information characterizes the declaration of identifiers and the portions of the program where it is allowed to use each identifier. Different languages have different scopes for declarations. For example, in FORTRAN, the scope of a name is a single subroutine, whereas in ALGOL, the scope of a name is the section or procedure in which it is declared. Thus, the same identifier may be declared several times as distinct names, with different attributes, and with different intended storage locations. The symbol table is thus responsible for keeping different declaration of the same identifier distinct.

To make distinction among the declarations, a unique number is assigned to each program element that in return may have its own local data. Semantic rules associated with productions that can recognize the beginning and ending of a subprogram are used to compute the number of currently active subprograms.

There are mainly two semantic rules regarding the scope of an identifier.

- Each identifier can only be used within its scope.
- Two or more identifiers with same name and are of same kind cannot be declared within the same lexical scope.

The scope declaration of variables, functions, labels and objects within a program is shown here.

Scope of variables in statement blocks:

```
{int x;
   ...                  <-- Scope of variable x
      {int y;
         ...            <-- Scope of variable y
      }
   ...
}
```

- **Scope of formal arguments of functions:**

```
int mul (int n) {
   ...                  <-- Scope of argument n
}
```

- **Scope of labels:**

```
void jumper () {
   ...goto sim;
   ...
   sim++;               <-- Scope of label sim
   ...goto sim;
   ...
}
```

- **Scope in class declaration (scope of declaration):** The portion of the program in which a declaration can be applied is called the scope of that declaration. In a procedure a name is said to be local to the procedure if it is in the scope of declaration within the procedure, otherwise the name is said to be non-local.

Scope of object fields and methods:

```
class X {
   public:
      void A()
      {
         m = 1;
      }
   private:
      int m;
      . . .
}
```

Scope of variable m and method A

8. Differentiate between lexical scope and dynamic scope.

Ans: The differences between lexical scope and dynamic scope are given in Table 10.1.

Table 10.1 Difference between lexical and dynamic scope

Lexical Scope	Dynamic Scope
• The binding of name occurrences to declarations is done statically at compile time. • The structure of the program defines the binding of variables. • A free variable in a procedure gets its value from the environment in which the procedure is defined.	• The binding of name occurrences to declarations is done dynamically at run time. • The binding of variables is defined by the flow of control at the run time. • A free variable gets its value from where the procedure is called.

9. Explain error detection and recovery in lexical phase, syntactic phase, and semantic phase.

Ans: The classification of errors is given in Figure 10.8.

These errors should be detected during different phases of compiler. Error detection and recovery is one of the main tasks of a compiler. The compiler scans and compiles the entire program, and errors detected during scanning need to be recovered as soon as they are detected.

Usually, most of the errors are encountered during syntax and semantic analysis phase. Every phase of a compiler expects the input to be in particular format, and an error is returned by the

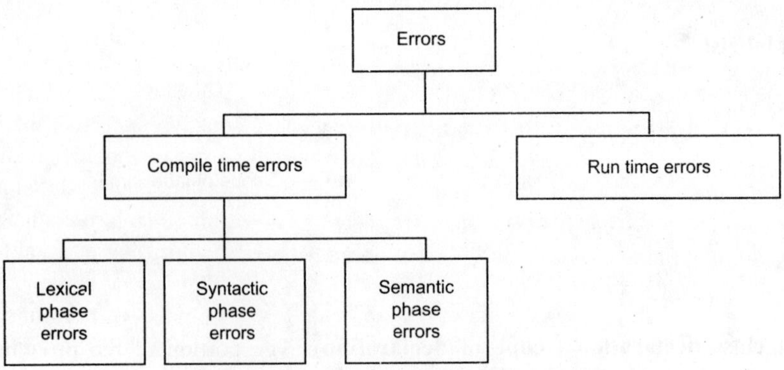

Figure 10.8 Classification of Errors

compiler whenever the input is not in the required format. On detection of an error, the compiler scans some of the tokens ahead of the point of error occurrence. A compiler is said to have better error-detection capability if it needs to scan only a few numbers of tokens ahead of the point of error occurrence.

A good error detection scheme reports errors in an intelligible manner and should possess the following properties.

- The error message should be easily understandable.
- The error message should be produced in terms of original source program and not in any internal representation of the source program. For example, each error message should have a line number of the source program associated with it.
- The error message should be specific and properly localize the error. For example, an error message should be like, "`A` is not declared in function `sum`" and not just "missing declaration".
- The same error message should not be produced again and again, that is, there is no redundancy in the error messages.

Error detection and recovery in lexical Phase: The errors where the remaining input characters do not form any token of the language are detected by the lexical phase of compiler. Typical lexical phase errors are spelling errors, appearance of illegal characters and exceeding length of identifier or numeric constant.

Once an error is detected, the lexical analyzer calls an error recovery routine. The simplest error recovery routine skips the erroneous characters in the input until the lexical analyzer finds a synchronizing token. But this scheme causes the parser to have a deletion error, which would result in several difficulties for the syntax analysis and for the rest of the phases.

The ability of lexical analyzer to recover from errors can be improved by making a list of legitimate tokens (in the current context) which are accessible to the error recovery routine. With the help of this list, the error recovery routine can decide whether the remaining input characters match with a synchronizing token and can be treated as that token.

Error detection and recovery in syntactic phase: The errors where the token stream violates the syntax of the language and the parser does not find any valid move from its current configuration are detected during the syntactic phase of the compiler. The `LL(1)` and `LR(1)` parsers have valid prefix property capability, that is, they report an error as soon as they read an input character which is not a valid continuation of the previous input prefix. In this way, these parsers reduce the amount of erroneous output to be passed to next phases of the compiler.

To recover from these errors, *panic mode recovery* scheme or *phrase level recovery* scheme (discussed in chapter 04) can be used.

Error detection and recovery in semantic phase: The language constructs that have the right syntactic structure but have no meaning to the operation involved are detected during semantic analysis phase. Undeclared names, type incompatibilities and mismatching of actual arguments with formal arguments are the main causes of semantic errors. When an undeclared name is encountered first time, a symbol table entry is created for that name with an attribute that is suitable to the current context.

For example, if semantic phase detects an error like "missing declaration of `A` in function `sum`", then a symbol table entry is created for `A` with an attribute that is suitable to the current context. To indicate that an attribute has been added to recover from an error and not in response to the declaration of `A`, a flag is set in the `A` symbol table record.

Multiple-Choice Questions

1. Which of the following is not true in context of a symbol table?
   (a) It is a compile time data structure.
   (b) It maps name into declarations.
   (c) It does not help in error detection and recovery.
   (d) It contains formal parameter list and return type of each function and procedure.
2. The information in the symbol table is entered during ________.
   (a) Lexical analysis
   (b) Syntax analysis
   (c) Both (a) and (b)
   (d) None of these
3. Which of these operations can be performed on a symbol table?
   (a) Insert
   (b) Lookup
   (c) begin_scope and end_scope
   (d) All of these
4. Which of the following data structure is not used to implement symbol tables?
   (a) Linear list
   (b) Hash table
   (c) Binary search tree
   (d) AVL tree
5. Which of the following is not true for scope representation in symbol table?
   (a) Declarations have same scope in different languages.
   (b) The scope of a name is a single subroutine in FORTRAN.
   (c) Symbol table keeps different declaration of the same identifier distinct.
   (d) In ALGOL, the scope of a name is the section or procedure in which it declared.
6. Which of the following is not true for error detection and recovery?
   (a) Error detection and recovery is the main task of the compiler.
   (b) Most of the errors are detected during lexical phase.
   (c) A compiler returns an error, if the input is not in the required format.
   (d) None of these

Answers

1. (c) 2. (b) 3. (c) 4. (b) 5. (a) 6. (c) 7. (b)

11

Code Optimization and Code Generation

1. Explain the various issues in the design of code generator.

Or

Discuss the various factors affecting the code generation process.

Ans: The various factors that affect the code generation process are as follows:

- **Input:** The intermediate code produced by the intermediate code generator or code optimizer of the compiler is given as input to the code generator. At the time of code generation, the source program is assumed to be scanned, parsed, and translated into a relatively low-level intermediate representation. Type conversion operators are assumed to be inserted wherever required, and that semantic errors have also been detected. The code generation phase, therefore, proceeds on the assumption that the input to the code generator is free from errors. We also assume that the operators, data types, and the addressing modes appearing in the intermediate representation can be directly mapped to the target machine representation. If such straightforward mappings exist, then the code generation is simple, otherwise a significant amount of translation effort is required.
- **Structure of target code:** The efficient construction of a code generator depends mainly on the structure of the target code which further depends on the instruction-set architecture of the target machine. RISC (reduced instruction set computer) and CISC (complex instruction set computer) are the two most common target machine architectures. The target program code may be absolute machine language code, relocatable machine language code, or assembly language code.
    - If the target program code is absolute machine language code, then it can be placed in a fixed memory location and can be executed immediately. The fixed location of program variables and code makes the absolute code generation relatively easier.
    - If the target program code is relocatable machine language code (also known as **object module**), then the code generation becomes a bit difficult as relocatable code may or may not be supported by the underlying hardware. In case the target machine does not support relocation automatically, it is the responsibility of compiler to explicitly insert the code for ensuring smooth relocation. However, producing a relocatable code requires subprograms to

be compiled separately. After compilation, all the relocatable object modules can be linked together and loaded for execution by a linking loader.

- If the output is assembly language program, then it can be converted into an executable version by an assembler. In this case, the code generation can be made simpler by utilizing the features of assembler. That is, we can generate symbolic instruction code and use the macro facilities of the assembler to help the code generation process.

- **Selection of instruction:** The nature of the instruction set of the target machine is an important factor to determine the complexity of instruction selection. The uniformity and completeness of the instruction set, instruction speed, and machine idioms are the important factors that are to be considered. If we are not concerned with the efficiency of the target program, then instruction selection becomes easier and straightforward. The two important factors that determine the quality of the generated code are its speed and size.
For example, the three-address statement of the form,

```
A = B + C
X = A + Y
```

can be translated into a code sequence as given below:

```
LD    R0,B
ADD   R0,R0,C
ST    A,R0
LD    R0,A
ADD   R0,R0,Y
ST    X,R0
```

The main drawback of this statement by statement code generation is that it produces redundant load and store statements. For example, the fourth step in the above code is redundant as the value that has been stored just before is loaded again. If the target machine provides a rich set of instructions then there will be several ways of implementing a given instruction. For example, if the target machine has an increment instruction, `X = X+1`, then instead of multiple load and store instructions, we can have simple instruction `INC X`. Note that deciding which machine-code sequence is suitable for a given set of three-address instructions may require knowledge about the context in which those instructions appear.

- **Allocation of registers:** Assigning the values to the registers is the key problem during code generation. So, generation of a good code requires the efficient utilization of registers. In general, the utilization of registers is subdivided into two phases, namely, register allocation and register assignment. **Register allocation** is the process of selecting a set of variables that will reside in CPU registers. **Register assignment** refers to the assignment of a variable to a specific register. Determining the optimal assignment of registers to variables even with single register values is difficult because the allocation problem is NP-complete. In certain machines, even/odd register pairs are required for some operands and results which make the problem further complicated. In integer multiplication, the multiplicand is placed in the odd register, however, the multiplier can be placed in any other single register, and the product (result) is placed in the entire even/odd register pair. Register allocation becomes a nontrivial task because of these architecture-specific issues.

- **Evaluation order:** The performance of the target code is greatly affected by the order in which computations are performed. For some computation order, only a fewer registers are required to hold the intermediate results. Hence, deciding the optimal computation order is again difficult since the problem is NP–complete. The problem can be avoided initially by generating the code for the three-address statements in the same order as that of produced by the intermediate code generator.

2. Define basic block.

Ans: A **basic block** is a sequence of consecutive three-address statements in which the flow of control enters only from the first statement of the basic block and once entered, the statements of the block are executed without branching, halting, looping or jumping except at the last statement. The control will leave the block only from the last statement of the block. For example, consider the following statements.

t_1: = X * Y
t_2: = 5 * t_1
t_3: = T_1 * t_2

In the above sequence of statements, the control enters only from the first statement, t_1: = X * Y. The second and third statements are executed sequentially without any looping or branching and the control leaves the block from the last statement. Hence, the above statements form a basic block.

3. Write the steps for constructing leaders in a basic block.

Or

How can you find leaders in basic blocks?

Ans: The first statement in the basic block is known as the **leader**. The rules for finding leaders are as follows:

(i) The first statement in the intermediate code is leader.

(ii) The target statement of a conditional and unconditional jump is a leader.

(iii) The inmediate statement following an unconditional or conditional jump is a leader.

4. Write an algorithm for partitioning of three-address instructions into a basic block. Give an example also.

Ans: A sequence of three-address instructions is taken as input and the following steps are performed to partition the three-address instructions into basic blocks:

Step 1: Determine the set of leaders.

Step 2: Construct the basic block for each leader that consists of the leader and all the instructions till the next leader (excluding the next leader) or the end of the program.

The instructions that are not included in a block are not executed and may be removed, if desired.

For example, consider the following code segment that computes a dot product between two integer arrays X and Y.

```
begin
    PRODUCT: = 0
    j: = 1
    do
    begin
        PRODUCT: = PRODUCT + X[j] * Y[j]
        j: = j + 1
    end
    while j <= 20
```

The corresponding three-address code for the above code segment is given as follows:

```
1.  PRODUCT: = 0
2.  j: = 1
3.  t1: = 4 * j /* assuming that the elements of integer array
    take 4 bytes*/
4.  t2: = X[t1] /* computing X[j] */
5.  t3: = 4 * j
6.  t4: = Y[t2] /* computing Y[j] */
7.  t5: = t2 * t4 /* computing
    X[j] * Y[j] */
8.  t6: = PRODUCT + t5
9.  PRODUCT: = t6
10. t7: = j + 1
11. j: = t7
12. if j <= 20 goto (3)
```

```
1.  PRODUCT:= 0          } Block B1
2.  j: = 1
3.  t1: = 4 * j          )
4.  t2: = X[t1]          |
5.  t3: = 4 * j          |
6.  t4: = Y[t2]          |
7.  t5: = t2 * t4        } Block B2
8.  t6:= PRODUCT + t5    |
9.  PRODUCT: = t6        |
10. t7: = j + 1          |
11. j: = t7              |
12. if j <= 20 goto (3)  )
```

Figure 11.1 Basic Blocks

Now, we can determine the basic blocks of the above three-address code by following the previous algorithm. Considering the rules for finding the leaders, according to rule (i) statement (1) is a leader. According to rule (ii), statement (3) is a leader. According to rule (iii), the statement following the 12th statement, if any, is a leader. Hence, the statements (1) and (2) form the first basic block and the rest of the program starting with statement (3) forms the second basic block as shown in Figure 11.1.

5. Explain the role of flow graph in basic blocks.

Ans: A **flow graph** is a directed graph that represents flow of control between the basic blocks. The basic block represents the nodes of the graph, and the edges define the control transfer. The flow graph for the program given in previous question is given in Figure 11.2.

If block B_2 immediately follows block B_1 then, there is an edge from block B_1 to B_2. B_1 is said to be the **predecessor** of B_2 and B_2 is said to be the **successor** of B_1 if, any of the following two conditions are satisfied.

- There is an unconditional or conditional jump from the last instruction of the block B_1 to the starting instruction of block B_2.
- B_1 does not end in an unconditional jump and block B_1 is immediately followed by block B_2.

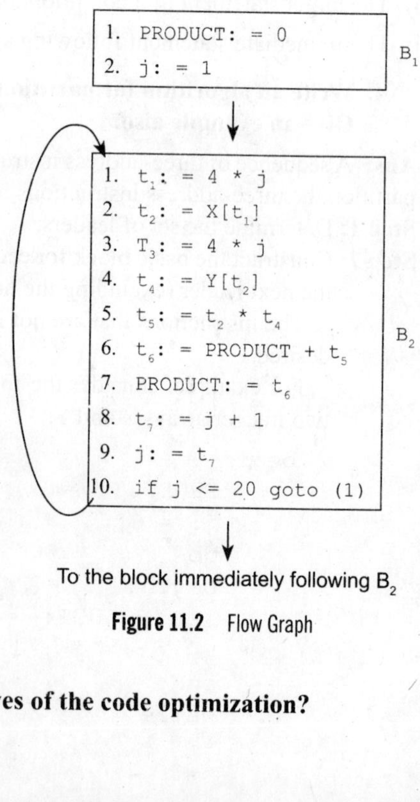

Figure 11.2 Flow Graph

6. Explain code optimization. What are the objectives of the code optimization?

Ans: Code optimization is an attempt of compiler to produce a better object code (target code) with high execution efficiency than the input source program. In some cases, code optimization can be so simple that it can be carried out without much difficulty. However, in some other cases, it may require a complete analysis of the program. Code optimization may require various transformation of the source program. These transformations should be carried out in such a way that all the translations of a source program are semantic equivalent, as well as the algorithm should not be modified in any case. The efficiency and effectiveness of a code optimization technique is determined by the time and space required by the compiler to produce the target program. Code optimization can be machine dependent or machine independent (discussed in Question 15).

Objectives of the Code Optimization:

- Production of target program with high execution efficiency.
- Reduction of the space occupied by the program.
- Time efficient program that takes lesser compilation time.

7. Write a short note on optimizing transformations.

Ans: Optimizing transformations are of two types, namely, *local* and *global*. Local optimization performed on each basic block and global optimization is applied over the large segments of a progra consisting of loops or procedure/function. Local optimization involves transforming a basic block in a DAG and global optimization involves the data flow analysis. Though local optimization requires le amount of code analysis, however, it does not allow all kinds of code optimizations (for example, lo optimizations cannot be performed locally). However, local optimization can be merged with the initi phase of global optimization to simplify the global optimization.

8. What is constant folding?

Ans: Constant folding is used for code optimization. It evaluates the constant expressions duri compile time and replaces the expressions by their values. For example, consider the followi statements:

```
X: = 5 + 2
A: = X + 2
```

After constant folding, the statements can be written as follows:

```
X: = 7
A: = 9
```

9. Explain loop optimization.

Ans: Loop optimization is a technique in which inner loops are taken into consideration the code optimization. Only the inner loops are considered because a large amount of time is tak during the execution of these inner loops. The various loop optimization techniques are *loop-invari expression elimination*, *induction variable elimination*, *strength reduction*, *loop unrolling* and *lc fusion*.

- **Loop-invariant expression elimination:** An expression is said to be *loop-invariant* express if it produces the same result each time the loop is executed. During loop-invariant express elimination, we eliminate all such expressions by placing them at the entry point of the loop. I example, consider the following code segment:

```
if (i > Min + 2)
{
    sum = sum + x[i];
}
```

In this code segment, the expression `Min + 2` is evaluated each time the loop is executed, however, it always produces the same result irrespective of the iteration of the loop. Thus, we can place this expression at the entry point of the loop as follows:

```
n = Min + 2
if (i > n)
{
    sum = sum + x[i];
}
```

Since in loop-invariant expression elimination, the expression from inside the loop is moved outside the loop, this method is also known as **loop-invariant code motion**.

- **Induction variable elimination:** A variable is said to be *induction variable* if its value gets incremented or decremented by some constant every time the loop is executed. For example, consider the following `for loop` statement:

```
for (j = 1;j <= 10; j++)
```

Here, the value of `j` is incremented every time the loop is executed. Hence, `j` is an induction variable. If there are more than one induction variables in a loop then, it is possible to get rid of all but one. This process is known as **induction variable elimination**.

- **Strength reduction:** Replacing an expensive operation with an equivalent cheaper operation is called **strength reduction**. For example, the `*` operator can be replaced by a lower strength operator +. Consider the following code segment:

```
for (j = 1; j <= 10; j++)
{
    ...
    cnt = j * 5;
    ...
}
```

After strength reduction, the code can be written as follows:

```
temp = 5;
for (j = 1; j <= 10; j++)
{
    ...
    cnt = temp;
    temp = temp + 5;
    ...
}
```

- **Loop unrolling:** The number of jumps can be reduced by replicating the body of the loop if the number of iterations is found to be constant (that is, the number of iterations is known at compile time). For example, consider the following code segment:

```
int j = 1;
while (j <= 50)
{
    X[j] = 0;
    j = j + 1;
}
```

This code segment performs the test 50 times. The number of tests can be reduced to 25 by replicating the code inside the body of the loop as follows:

```
int j = 1;
while (j <= 50)
{
    X[j] = 0;
    j = j + 1;
    X[j] = 0;
    j = j + 1;
}
```

The main problem with loop unrolling is that if the body of the loop is big, then unrolling may increase the code size, which in turn, may affect the system performance.

- **Loop fusion:** It is also known as **loop jamming** in which the bodies of the two loops are merged together to form a single loop provided that they do not make any references to each other. For example, consider the following statements:

```
int i,j;
for(i = 1;i <= n; i++)
    A[i] = B[i];
for(j = 1;j <= n; j++)
    C[j] = A[j];
```

After loop fusion, this code can be written as follows:

```
int i,j;
for(k = 1; k <= n; k++)
{
    A[k] = B[k];
    C[k] = A[k];
}
```

10. Define DAG (Directed Acyclic Graph). Discuss the construction of DAG of a given basic block.

Ans: A **DAG** is a directed acyclic graph that is used to represent the basic blocks and to implement transformations on them. It represents the way in which the value computed by each statement in a basic block is used in the subsequent statements in the block. Every node in a flow graph can be represented by a DAG. Each node of a DAG is associated with a label. The labels are assigned by using these rules:

- The leaf nodes are labeled by unique identifiers which can be either constants or variable names. The initial values of names are represented by the leaf nodes, and hence they are subscripted with 0 in order to avoid confusion with labels denoting current values of names.
- An operator symbol is used to label the interior nodes.
- Nodes are also labeled with an extra set of identifiers where the interior nodes represent the computed values and the identifier labeling that node contains the computed value.

The main difference between a flow graph and a DAG is that a flow graph consists of several nodes where each node stands for a basic block, whereas a DAG can be constructed for each node (or the basic block) in the flow graph.

Construction of DAG

While constructing a DAG, we consider a function `node(identifier)` which returns the most recently created node associated with an identifier. Assume that there are no nodes initially and `node( )` is undefined for all arguments. Let the three-address statements be either

(i) `X: = Y op Z or`
(ii) `X: = op Y or`
(iii) `X: = Y`

The steps followed for the construction of DAG are as follows:

1. Create a leaf labeled `Y` if `node(Y)` is undefined and let that node be `node(Y)`. If `node(Z)` is undefined for three-address statement (i) then, create a leaf labeled `Z` and let it be `node(Z)`.
2. Determine if there is any node labeled `op`, where `node(Y)` as its left child and `node(Z)` as its right child for three-address statement (i). If such a node is not found, then create such a node and let it be `n`. For three-address statement (ii), determine if there is any node labeled `op`, whose only child is `node(Z)`. If such a node is not found then, create a node and let it be `n`. Similarly, create a node `n` for `node(Y)` for three-address statement (iii).
3. Delete `X` from the list of attached identifiers for `node(X)`. Append `X` to the list of attached identifiers for node `n` created or found in step 2 and set `node(X)` to `n`.

For example, consider the block B_2 shown in Figure 11.2. For the first statement, t_1`: = 4 * j`, leaves labeled `4` and j_0 are created. In the next step, a node labeled `*` is created and t_1 is attached to it as an identifier. The DAG representation is shown in Figure 11.3(a).

For the second statement t_2`: = X[`t_1`]`, a new leaf labeled `X` is created. Since, we have already created `node(`t_1`)` in the previous step, we will not create a new node t_1. However, we create a new node for `[]`, and attach `X` and t_1 as its child nodes. Now, the third statement t_3`: = 4 * j` is same as that of first statement, therefore, we will not create any new node; rather we give the existing `*` node an additional label t_3. The DAG representation for this is shown in Figure 11.3(b).

For the fourth statement, t_4`: = Y[`t_3`]` we create a node `[]` and attach `Y` as its left child node. The corresponding DAG representation is shown in Figure 11.3(c). For the fifth statement t_5`: =` t_2 `+` t_4, we create a new node + and attach the already created nodes labeled t_4 and t_2 as its left and right child respectively. The resultant DAG is shown in Figure 11.3(d).

For the sixth statement, we create a new node labeled +, and attach a leaf labeled $PRODUCT_0$ as its left child. The already created `node(*)` is attached as its right child. For the seventh statement, `PRODUCT: =` t_6, we assign the additional label `PRODUCT` to the existing + node. The resultant DAG is shown in Figure 11.3(e).

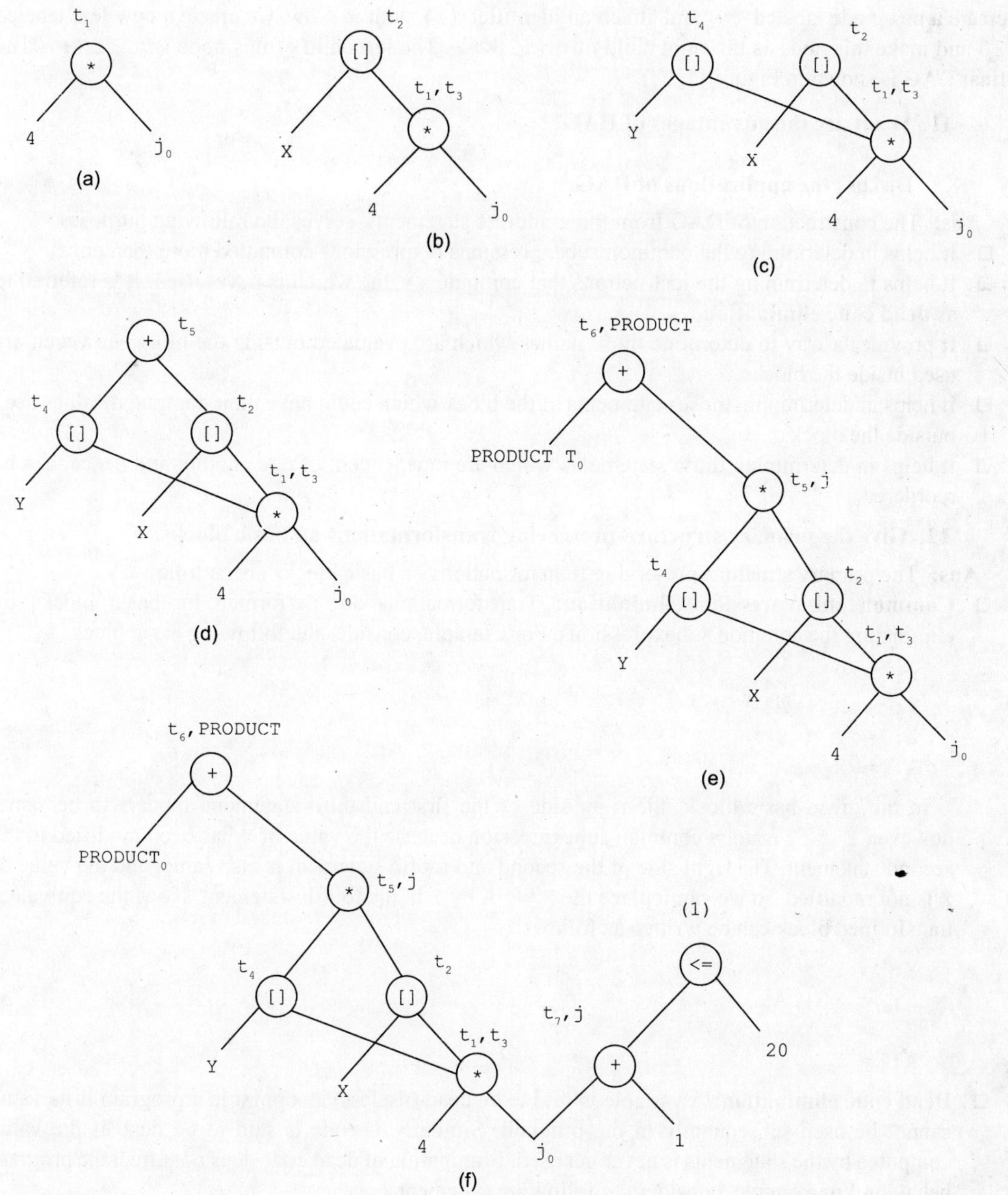

Figure 11.3 Step-by-Step Construction of DAG

For the eighth statement t_7 : = j + 1, we create a new node labeled +, and make j_0 as its left child. Now, we create a new leaf labeled 1 and make this leaf as its right child. For the ninth statement, we will not create any new node; rather we give node + the additional label j. Finally, for the last statement we create a new node labeled <= and attach an identifier (1) with it. Now, we create a new leaf labeled 20 and make this node as the right child of node(<=). The left child of this node is node(+). The final DAG is shown in Figure 11.3(f).

11. What are the advantages of DAG?

Or

Discuss the applications of DAG.

Ans: The construction of DAG from three-address statements, serves the following purposes:

- It helps in determining the common subexpressions (expressions computed more than once).
- It helps in determining the instructions that compute a value which is never used. It is referred to as **dead code elimination**.
- It provides a way to determine those names which are evaluated outside the block, however, are used inside the blocks.
- It helps in determining those statements of the block which could have their computed values used outside the block.
- It helps in determining those statements which are independent of one another and hence, can be reordered.

12. Give the primary structure-preserving transformations on basic blocks.

Ans: The primary structure-preserving transformations on basic blocks are as follows:

- **Common subexpression elimination:** Transformations are performed on basic blocks by eliminating the common subexpressions. For example, consider the following basic block:

```
X: = Y * Z
Y: = X + A
Z: = Y * Z
A: = X + A
```

In the given basic block, the right side of the first and third statement appears to be same, however, Y * Z is not a common subexpression because the value of Y has been modified in the second statement. The right side of the second and fourth statement is also same, and the value of X is not modified, so we can replace the X + A by Y in the fourth statement. Now, the equivalent transformed block can be written as follows:

```
X: = Y * Z
Y: = X + A
Z: = Y * Z
A: = Y
```

- **Dead code elimination:** A variable is said to be dead (useless) at a point in a program if its value cannot be used subsequently in the program. Similarly, a code is said to be dead if the value computed by the statements is never get used. Elimination of dead code does not affect the program behavior. For example, consider the following statements:

```
flag: = false
If (flag) print some information
```

Here, the print statement is dead as the value of flag is always *false* and hence the control never reaches to the print statement. Thus, the complete if statement (test and the print operation) can be eliminated easily from the object code.

- **Renaming temporary variables:** A statement of the form t_1: = X + Y, where t_1 is a temporary variable can be changed to a statement t_2:= X + Y where, t_2 is a new temporary variable. Thus, all the instances of t_1 can be changed to t_2 without affecting the value of the block.
- **Interchange of statements:** Two statements can be interchanged in the object code if they make no references to each other, and their order of execution does not affect the value of the block. For example, consider the following statements:

```
t1: = A + B
t2: = X + Y
```

  If neither X nor Y is t_1 and neither A nor B is t_2, then the two statements can be interchanged without affecting the value of the block.
- **Code motion:** Moving the code from one part of the program to another so that the resultant program is equivalent to the original one is referred to as **code motion**. Code motion is performed to reduce the size of the program and to reduce the execution frequency of the code which is moved. For example, consider the following code segment:

```
if (x < y)
    result: = x * 2
else
    result: = x * 2 + 50
```

  In this code segment, the subexpression x * 2 is evaluated twice, thus, it can be moved before if statement as shown below:

```
temp: = x * 2
if (x < y)
    result: = temp
else
    result: = temp + 50
```

- **Variable propagation:** In variable propagation, a variable is replaced by another variable having identical value. For example, consider the following statements:

```
X: = Y
A: = X * B
C: = Y * B
```

  The statement X: = Y specifies that the values of X and Y are equal. Since the value of X or Y is not modified further, the second statement can hence be written as A: = Y + B by propagating the variable Y to it. This propagation makes Y * B as a common subexpression in the last two statements, and hence possibly evaluated only once.
- **Algebraic transformations:** In algebraic transformations, the algebraic identities are used to optimize the code. Some of the common algebraic identities are given below:

```
A + 0 = A    (Additive identity)
A * 1 = A    (Multiplicative identity)
A * 0 = 0    (Multiplication with 0)
```

  These identities are generally applied on a single intermediate code statement. For example, consider the following statements:

```
Y: = X + 0
Y: = X * 1
Y: = X * 0
```

After algebraic transformations, the expensive addition and multiplication operations involved in these statements can be replaced by cheaper assignment operations as given below:

```
Y: = X
Y: = X
Y: = 0
```

- **Induction variables and strength reduction:** Refer Question 9

13. Discuss in detail about a simple code generator with the appropriate algorithm.

Or

Explain code generation phase with simple code generation algorithm.

Ans: A simple code generator generates the target code for the three-address statements. The main issue during code generation is the utilization of registers since the number of registers available is limited. The code generation algorithm takes the sequence of three-address statements as input and assumes that for each operator, there exists a corresponding operator in target language. The machine code instruction takes the required operands in registers, performs the operation and stores the result in a register. Register and address descriptors are used to keep track of register contents and addresses.

- **Register descriptors** are used to keep track of the contents of each register at a given point of time. Initially, we assume that a register descriptor shows that all registers are empty and as the code generation proceeds, each register holds the value of zero or more names at some point.
- **Address descriptors** are used to trace the location of the current value of the name at run time. The location may be memory address, register, or a stack location, and this information can be stored in the symbol table to determine the accessing method for a name.

  The code generation algorithm for a three-address statement `X: = Y op Z` is given below:

  1. Call `getreg()` to obtain the location `L` where the result of `Y op Z` is to be stored. `L` can be a register or a memory location.
  2. Determine the current location of `Y` by consulting the address descriptor for `Y` and let it be `Y'`. If both the memory and register contains the value of `Y`, then prefer the register for `Y'`. If the value is not present in `L`, then generate an instruction `MOV Y', L`.
  3. Determine the current location of `Z`, say, `Z'` and generate the instruction `OP Z', L`. In this case also, if both the memory and the register hold the value of `Z`, then prefer the register. Update the address descriptor of `X` to indicate that `X` is in `L` and if `L` is a register then, update its descriptor indicating that it holds the value of `X`. Delete `X` from other register descriptors.
  4. If the current values of `Y` and/or `Z` are in registers, and if they have no further uses and are not live at the end of the block, then alter the register descriptor. This alteration indicates that `Y` and/or `Z` will no longer be present in those registers after the execution of `X: = Y op Z`.

  For the three-address statement `X: = OP Y`, the steps are analogous to the above steps. However, for the three-address statement of the form `X: = Y`, some modifications are required as discussed here.

- If `Y` is in a register then the register and address descriptors are altered to record that from now onwards the value of `X` is found only in the register that holds the value of `Y`.

- If Y is in the memory, the getreg() function is used to determine a register in which the value of Y is to be loaded, and that register is now made as the location of X.

Thus, the instruction of the form X: = Y could cause the register to hold the value of two or more variables simultaneously.

Implementing the Function getreg

For the three-address statement, X: = Y OP Z, the function getreg() returns a location L as follows:

1. If Y is in a register and it is not live and has no next uses after the execution of three-address statement, then return the register of Y for L. The address descriptor of Y is then updated to indicate that Y is no more present in L.
2. If Y is not in a register, then return an empty register for L (if exists).
3. If X is to be used further in the block or OP is the operator that requires a register, then find an occupied register R_0 that may contain one or more values. For each variable in R_0, issue a MOV R_0, M instruction to store the value of R_0 into a memory location M. Then, update the address descriptor for M, and return R_0. Though there are several ways to choose a suitable occupied register, the simplest way is to choose the one whose data values are to be referenced furthest in the future.
4. If X is not used in the block or an occupied register could not be found, then select the memory location of X as L.

14. Explain the peephole optimization and its characteristics.

Ans: Peephole optimization is an efficient technique for optimizing either the target code or the intermediate code. In this technique, a small portion of the code (known as **peephole**) is taken into consideration and optimization is done by replacing the code by the equivalent code with shorter or faster sequence of execution. The statements within the peephole need not be contiguous, although some of the implementations require the statements to be contiguous. Each improvement in the code may explore the opportunities for some other improvements. So, multiple review of the code is necessary to get maximum benefit from the peephole optimization. Some characteristics of the peephole optimization are: *redundant- instruction elimination, unreachable code elimination, flow of control optimizations, strength reduction* and *use of machine idioms.*

- **Redundant-instruction elimination:** Consider the following instructions:

```
MOV R0, X
MOV X, R0
```

The second instruction can be deleted since the first instruction ensures that the value of X is already loaded into register R_0. However, it cannot be deleted in a situation when, it has a label which makes it difficult to identify that whether the first instruction is always executed before the second. To ensure that this kind of transformation in the target code would be safe, the two instructions must be in the same basic block.

- **Unreachable code elimination:** Removing an unlabeled instruction that immediately follows an unconditional jump is possible. This process eliminates a sequence of instructions when repeated. Consider the following intermediate code representation:

```
if error == 1 goto L1
goto L2
```

```
L1: Print error information
L2:
```

Here, the code is executed only if the variable `error` is equal to 1. Peephole optimization allows the elimination of jumps over jumps. Hence, the above code is replaced as follows irrespective of the value of the variable `debug`.

```
if error ! = 1 goto L2
Print error information
L2:
```

Now, if the value of the variable is set to `0`, the code becomes:

```
if 0! = 1 goto L2
Print error information
L2:
```

Here, the first statement always evaluates to *true*. Hence, the statement printing the error information is unreachable and can be eliminated.

- **Flow of control optimizations:** The peephole optimization helps to eliminate the unnecessary jumps in the intermediate code. For example, consider the following code sequence:

```
goto L1
...
L1: goto L2
```

This sequence can be replaced by

```
goto L2
...
L1: goto L2
```

Now, if there are no jumps to `L1` and the statement `L1: goto L2` is preceded by an unconditional jump, then this statement can be eliminated. Similarly, consider the following code sequence:

```
if (x < y) goto L1
...
L1: goto L2
```

This sequence can be rewritten as follows:

```
if (x < y) goto L2
...
L1: goto L2
```

- **Strength reduction:** Peephole optimization also allows applying strength reduction transformations to replace expensive operations by the equivalent cheaper ones. For example, the expression X^2 can be replaced by an equivalent cheaper expression `X * X`.
- **Use of machine idioms:** Some target machines provide hardware instructions to implement certain operations in a better and efficient way. Thus, identifying the situations that permit the use of hardware instructions to implement certain operations may reduce the execution time significantly. For example, some machines provide auto-increment and auto-decrement addressing modes, which add and subtract one respectively from an operand. These modes can be used while pushing or popping a stack, or for the statements of the form `X: = X + 1` or `X: = X -- 1`. These transformations greatly improve the quality of the code.

15. Explain the machine-dependent and machine-independent optimization.

Ans: Machine-dependent optimizations: An optimization is called *machine-dependent optimization* if it requires knowledge of the target machine to perform optimization. A machine-dependent optimization utilizes the target system registers more efficiently than the machine-independent optimization. Often, machine-dependent optimizations are local and are considered as most effective for the local machine as these optimizations best exploit the features of target platform.

Machine-independent optimizations: The optimization which is not intended for a target machine specific platform and optimizations can be carried out independently of the target machine is known as *machine-independent optimizations*. Machine-independent optimizations can be both local and global and operates on abstract programming concepts (like loops, objects, and structures).

16. Discuss the value number method for constructing a DAG. How does it help in code optimization?

Ans: While constructing a DAG, we check whether a node with given operator and given children exist or not. If such a node does not exist then we create a node with that given operator and given children. To determine the existence of a node instantly, we can use a hash table. This idea of using a hash table is named as **value number method** by Cocke and Schwartz [1970]. Basically, the nodes of a DAG are stored in the form of array of records, where each record in the array corresponds to a node in the DAG. Each record consists of several fields, where the first field is always an operation code, which indicates the label of the node. The leaves have one more field holding the lexical value which may be a symbol-table pointer or a constant. The interior nodes have two more fields, which indicate the left and right child nodes. An array representation of a DAG shown in Figure 11.4(a) is shown in Figure 11.4(b).

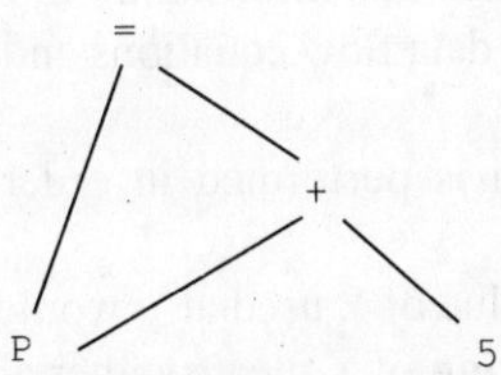

(a) An Example DAG

1	id	→ To entry for P	
2	num	5	
3	+	1	2
4	=	1	3
5	. . .		

(b) Array Representation

Figure 11.4 Value Number Method

In this array, the nodes are referred to by giving the integer index (called the value number) of the record for that node within the array. For instance, in Figure 11.4(b), the node labeled = has value number 4.

The value number method can also be used to implement certain optimizations based on algebraic laws (like *commutative, associative,* and *distributed laws*). For example, if we want to create a DAG node with its left child p and right child q, and operator *, we first check whether such a node exists by using value number method. As multiplication is commutative in nature, therefore, we also need to check the existence of a node labeled *, with its left child q and right child p.

The associative law can also be applied to improve the already generated code from a DAG. For example, consider the following statements:

```
P: = Q + R
S: = R + T + Q
```

The three-address code for these statements is given below:

```
1. t1: = Q + R
2. P: = t1
3. t2: = R + T
4. t3: = Q + t2
5. S: = t3
```

The DAG for this code is shown in Figure 11.5(a). If we assume that t_2 is not needed outside the block, then the DAG shown in Figure 11.5(a) can be changed to the one shown in Figure 11.5(b). Here, both the associative and commutative laws are used.

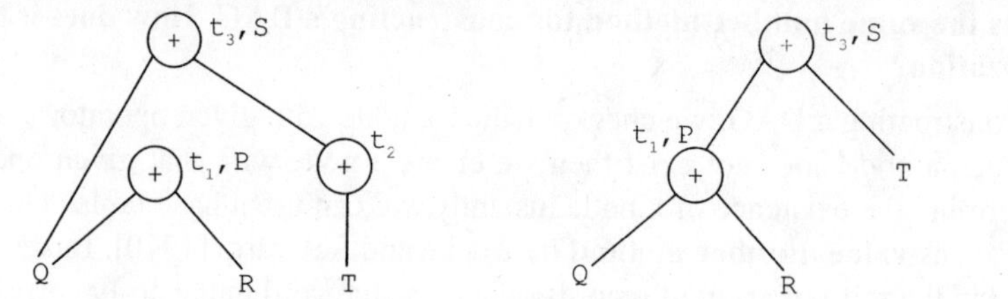

(a) DAG without Associative Law (b) DAG after Applying Associative Law

Figure 11.5 Use of Associative Law

17. What is global data flow analysis?

Ans: Global data flow analysis is a process to analyze how global data is processed and how analysis of the global data is useful in optimizations. Basically, the data flow analysis process collects the information about the program as a whole and then it distributes this information to each block of the flow graph. Data flow information is defined in terms of some data flow equations and then solving those equations to get the data flow information.

Ud-chaining: A global data flow analysis of the flow graph is performed in order to compute ud-chaining information. It answers the following question:

If a given identifier is used at point y, then at what point the value of X used at y would be defined? Here, the **use** of X means that X occurs as an operand, and **definition** of X means either an assignment to X or the reading of a value for X. A **point** refers to a position before and after any intermediate code statement. Within a graph, by assuming that all edges in the graph are traversable, we can say that, a definition of a variable X reaches a point y if there exists a path in flow graph from X's definition to y and no other definitions of X appear on the path.

Data flow equations: A data flow equation has the following form:

$$\text{Out[BB]} = \text{in[BB]} - \text{Kill[BB]} \cup \text{Gen[BB]} \quad (1)$$

where,

BB = Basic block

Gen[BB] = The set of all definitions generated in basic block BB.

Kill[BB] = The set of all definitions outside basic block BB that define the same variable as are defined in basic block BB.

$$\text{in[BB]} = \cup \text{ out[P]} \quad (2)$$

where,

P refers to the predecessor of BB.

The algorithm to find out the solutions of data flow equations is shown in Figure 11.6.

```
for each basic block BB do............................. (1)
begin
    in[BB] = Ø ........................................ (2)
    out[BB]: = Gen[BB] ................................ (3)
end
flag = true............................................ (4)
while(flag)do.......................................... (5)
begin
    flag = false ...................................... (6)
    for each block BB do .............................. (7)
    begin
          for each predecessor P of BB do ............. (8)
          in_new [BB] = in_new [BB] U out[P] .......... (9)
          if in_new ≠ in[BB] then ..................... (10)
          begin
            flag = true
            in[BB] = in_new [BB] ...................... (11)
            out[BB] = in[BB] - kill[BB] U gen [BB] .. (12)
          end
    end
end
```

Figure 11.6 Algorithm for Solving Data Flow Equations

18. Consider the following graph, and compute in and out of each block by using global data flow analysis.

Here, d_1, d_2, d_3, d_4, and d_5 are the definitions, and BB_1, BB_2, BB_3, BB_4, and BB_5 are the basic blocks.

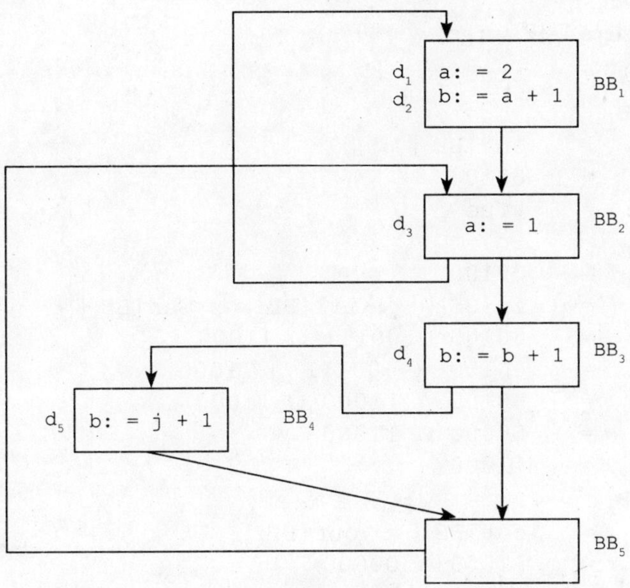

Ans: First, we need to compute `in` and `out` of each block, and for this we begin by computing `Gen` and `Kill` in BB_1. Both `a` and `b` are defined in block BB_1 hence, `Kill` contains all definitions of `a` and `b` outside the block BB_1.

$$Kill[BB_1] = \{d_3, d_4, d_5\}$$

As, d_1 and d_2 is the last definitions of their respective variables in BB_1, hence,

$$Gen[BB_1] = \{d_1, d_2\}$$

In BB_2, d_3 kills all definitions of `a` outside BB_2. Hence,

$$Kill[BB_2] = \{d_1\}$$
$$Gen[BB_2] = \{d_3\}$$

The complete list of `Gen`'s and `Kill`'s including their bit-vector representation is as follows:

Basic Block	Gen [BB]	Bit Vector	kill [BB]	Bit Vector
BB_1	$\{d_1, d_2\}$	11000	$\{d_3, d_4, d_5\}$	00111
BB_2	$\{d_3\}$	00100	$\{d_1\}$	10000
BB_3	$\{d_4\}$	00010	$\{d_2, d_5\}$	01001
BB_4	$\{d_5\}$	00001	$\{d_2, d_4\}$	01010
BB_5	Ø	00000	Ø	00000

Now, after performing steps 1–3 of algorithm given in Figure 11.6, we get the following initial iteration:

Basic Block	in [BB]	out [BB]
BB_1	00000	11000
BB_2	00000	00100
BB_3	00000	00010
BB_4	00000	00001
BB_5	00000	00000

After performing, steps 4–12, we get

Flag = true

For basic block BB_1

$$\begin{aligned}
in_{new} &= out[BB_2] \\
&= 00100 \\
Flag &= true \\
in[BB_1] &= in_{new} \\
&= 00100 \\
out[BB_1] &= in[BB_1] - Kill[BB_1] \cup Gen[BB_1] \\
&= 00100 - 00111 \cup 11000 \\
&= 00100 \wedge \neg 00111 \cup 11000 \\
&= 00100 \wedge 11000 \cup 11000 \\
&= 00000 \cup 11000 \\
&= 11000
\end{aligned}$$

For basic block BB_2

$$\begin{aligned}
in_{new} &= out[BB_1] \cup out[BB_5] \\
&= 11000 \cup 00000
\end{aligned}$$

$$
\begin{aligned}
& = 11000 \\
\text{Flag} & = \text{true} \\
\text{in}[BB_2] & = \text{in}_{new} \\
& = 11000 \\
\text{out}[BB_2] & = \text{in}[BB_2] - \text{Kill}[BB_2] \cup \text{Gen}[BB_2] \\
& = 11000 - 10000 \cup 00100 \\
& = 11000 \wedge \neg 10000 \cup 00100 \\
& = 11000 \wedge 01111 \cup 00100 \\
& = 01100
\end{aligned}
$$

For basic block BB_3

$$
\begin{aligned}
\text{in}_{new} & = \text{out}[BB_2] \\
& = 01100 \\
\text{Flag} & = \text{true} \\
\text{in}[BB_3] & = \text{in}_{new} \\
& = 01100 \\
\text{out}[BB_3] & = \text{in}[BB_3] - \text{Kill}[BB_3] \cup \text{Gen}[BB_3] \\
& = 01100 - 01001 \cup 00010 \\
& = 01100 \wedge \neg 01001 \cup 00010 \\
& = 01100 \wedge 10110 \cup 00010 \\
& = 00100 \cup 00010 \\
& = 00110
\end{aligned}
$$

For basic block BB_4

$$
\begin{aligned}
\text{in}_{new} & = \text{out}[BB_3] \\
& = 00110 \\
\text{in}[BB_4] & = \text{in}_{new} \\
& = 00110 \\
\text{Flag} & = \text{true} \\
\text{out}[BB_4] & = \text{in}[BB_4] - \text{Kill}[BB_4] \cup \text{Gen}[BB_4] \\
& = 00110 - 01010 \cup 00001 \\
& = 00110 \wedge \neg 01010 \cup 00001 \\
& = 00110 \wedge 10101 \cup 00001 \\
& = 00100 \cup 00001 \\
& = 00101
\end{aligned}
$$

For basic block BB_5

$$
\begin{aligned}
\text{in}_{new} & = \text{out}[BB_4] \cup \text{out}[BB_3] \\
& = 00101 \cup 00110 \\
& = 00111 \\
\text{Flag} & = \text{true} \\
\text{in}[BB_5] & = \text{in}_{new} \\
& = 00111 \\
\text{out}[BB_5] & = \text{in}[BB_5] - \text{Kill}[BB_5] \cup \text{Gen}[BB_5] \\
& = 00111 - 00000 \cup 00000 \\
& = 00111 \wedge \neg\, 00000 \cup 00000 \\
& = 00111 \wedge 11111 \cup 00000 \\
& = 00111 \cup 00000 \\
& = 00111
\end{aligned}
$$

Therefore, after pass 1, we get:

Basic Block	in [BB]	out [BB]
BB_1	00100	11000
BB_2	11000	01100
BB_3	01100	00110
BB_4	00110	00101
BB_5	00111	00111

Flag = true

For basic block BB_1

in_{new} = $out[BB_2]$
= 01100
Flag = true
$in[BB_1]$ = in_{new}
= 01100
$out[BB_1]$ = $in[BB_1] - Kill[BB_1] \cup Gen[BB_1]$
= $01100 - 00111 \cup 11000$
= $01100 \wedge \neg\ 00111 \cup 11000$
= $01100 \wedge 11000 \cup 11000$
= $01000 \cup 11000$
= 11000

For basic block BB_2

in_{new} = $out[BB_5] \cup out[BB_1]$
= $00111 \cup 11000$
= 11111
Flag = true
$in[BB_2]$ = in_{new}
= 11111
$out[BB_2]$ = $in[BB_2] - Kill[BB_2] \cup Gen[BB_2]$
= $11111 - 10000 \cup 00100$
= $11111 \wedge \neg\ 10000 \cup 00100$
= $11111 \wedge 01111 \cup 00100$
= $01111 \cup 00100$
= 01111

For basic block BB_3

in_{new} = $out[BB_2]$
= 01111
Flag = true
$in[BB_3]$ = in_{new}
= 01111
$out[BB_3]$ = $in[BB_3] - Kill[BB_3] \cup Gen[BB_3]$
= $01111 - 01001 \cup 00010$
= $01111 \wedge \neg\ 01001 \cup 00010$
= $01111 \wedge 10110 \cup 00010$

= 00110 ∪ 00010

= 00110

For basic block BB_4

in_{new} = out[BB_4]

= 00110

in[BB_4] = in_{new}

= 00110

out[BB_4] = in[BB_4] - Kill[BB_4] ∪ Gen[BB_4]

= 00110 - 01010 ∪ 00001

= 00110 ∧ ¬ 01010 ∪ 00001

= 00110 ∧ 10101 ∪ 00001

= 00100 ∪ 00001

= 00101

For basic block BB_5

in_{new} = out[BB_3] ∪ out[BB_4]

= 00110 ∪ 00101

= 00111

in[BB_5] = in_{new}

= 00111

out[BB_5] = in[BB_5] - Kill[BB_5] ∪ Gen[BB_5]

= 00111 - 00000 ∪ 00000

= 00111 ∧ ¬ 00000 ∪ 00000

= 00111 ∧ 11111 ∪ 00000

= 00111 ∪ 00000

= 00111

Therefore, after pass 2, we have:

Basic Block	in [BB]	out [BB]
BB_1	01100	11000
BB_2	11111	01111
BB_3	01111	00110
BB_4	00110	00101
BB_5	00111	00111

Flag = false

For basic block BB_1

in_{new} = out[BB_2]

= 01111

Flag = true

in[BB_2] = in_{new}

= 01111

out[BB_1] = in[BB_1] - Kill[BB_1] ∪ Gen[BB_1]

= 01111 - 00111 ∪ 11000

= 01111 ∧ ¬ 00111 ∪ 11000

= 01111 ∧ 11000 ∪ 11000

$$
\begin{aligned}
&= 01000 \cup 11000 \\
&= 11000
\end{aligned}
$$

For basic block BB_2

$$
\begin{aligned}
in_{new} &= out[BB_1] \cup out[BB_5] \\
&= 11000 \cup 00111 \\
&= 11000 \\
in[BB_2] &= in_{new} \\
&= 11111 \\
out[BB_2] &= in[BB_2] - Kill[BB_2] \cup Gen[BB_2] \\
&= 11111 - 10000 \cup 00100 \\
&= 11111 \land \neg\ 10000 \cup 00100 \\
&= 11111 \land 01111 \cup 00100 \\
&= 01111 \cup 00100 \\
&= 01111
\end{aligned}
$$

For basic bloc BB_3

$$
\begin{aligned}
in_{new} &= out[BB_3] \\
&= 01111 \\
in[BB_3] &= in_{new} \\
&= 01111 \\
out[BB_3] &= in[BB_3] - Kill[BB_3] \cup Gen[BB_3] \\
&= 01111 - 01001 \cup 00010 \\
&= 01111 \land \neg\ 01001 \cup 00010 \\
&= 01111 \land 10110 \cup 00010 \\
&= 00110 \cup 00010 \\
&= 00110
\end{aligned}
$$

For basic block BB_4

$$
\begin{aligned}
in_{new} &= out[BB_3] \\
&= 00110 \\
in[BB_4] &= in_{new} \\
&= 00110 \\
out[BB_4] &= in[BB_4] - Kill[BB_4] \cup Gen[BB_4] \\
&= 00100 - 01010 \cup 00001 \\
&= 00100 \land \neg\ 01010 \cup 00001 \\
&= 00100 \land 10101 \cup 00001 \\
&= 00100 \cup 00001 \\
&= 00101
\end{aligned}
$$

For basic block BB_5

$$
\begin{aligned}
in_{new} &= out[BB_3] \cup out[BB_4] \\
&= 00110 \cup 00101 \\
&= 00111 \\
in[BB_5] &= in_{new} \\
&= 00111 \\
out[BB_5] &= in[BB_5] - Kill[BB_5] \cup Gen[BB_5] \\
&= 00111 - 00000 \cup 00000 \\
&= 00111 \land \neg\ 00000 \cup 00000
\end{aligned}
$$

$$= 00111 \wedge 11111 \cup 00000$$
$$= 00111 \cup 00001$$
$$= 00111$$

Therefore, after pass 3, we get:

Basic Block	in [BB]	out [BB]
BB_1	01111	11000
BB_2	11111	01111
BB_3	01111	00110
BB_4	00110	00101
BB_5	00111	00111

In next pass, the value for `in` and `out` will be same, and hence these `in` and `out` values are final and correct.

Multiple-Choice Questions

1. An optimizing compiler __________.
   (a) is optimized to occupy less space
   (b) is optimized to take less time for execution
   (c) optimizes the code
   (d) None of these
2. A basic block can be analyzed by a __________.
   (a) DAG
   (b) Flow graph
   (c) Graph which may involve cycles
   (d) All of these
3. Reduction in strength means __________.
   (a) Replacing runtime computation
   (b) Removing loop-invariant computation
   (c) Removing common subexpressions
   (d) Replacing a costly operation by a cheaper one
4. Which of the following is not true for a DAG?
   (a) DAG cannot implement transformations on basic blocks.
   (b) The nodes of DAG correspond to the operations in the basic block
   (c) Each node of a DAG is associated with a label.
   (d) None of these
5. Which of the following comments about peephole optimization?
   (a) It is applied to a small part of the code.
   (b) It can be used to optimize intermediate code.
   (c) It can be applied to a portion of the code that is not contiguous.
   (d) All of these

6. A variable is said to be __________ if its value gets incremented or decremented every time by some constant.
   (a) Induction variable
   (b) Dead
   (c) Live
   (d) None of the above
7. __________ is the process of selecting a set of variables that will reside in CPU registers.
   (a) Register assignment
   (b) Register allocation
   (c) Instruction selection
   (d) None of these
8. Which of the following outputs can be converted into executable version by an assembler?
   (a) Absolute machine language
   (b) Relocatable machine language
   (c) Assembly language
   (d) None of the above
9. In __________ the bodies of the two loops are merged to form a single loop.
   (a) Loop unrolling
   (b) Strength reduction
   (c) Loop concatenation
   (d) Loop fusion
10. __________ are used to trace the location of the current value of the name at runtime.
    (a) Register descriptors
    (b) Address descriptors
    (c) Both (a) and (b)
    (d) None of these

Answers

1. (c) 2. (a) 3. (d) 4. (a) 5. (d) 6. (a) 7. (b) 8. (c) 9. (d) 10. (b)

Index

D

E

F

G

H

I

J

L

M

N

O

P

Q

R

S

T

U

V

Y